Canto is an imprint offering a range of
titles, classic and more recent, across a
broad spectrum of subject areas and
interests. History, literature, biography,
archaeology, politics, religion, psychology,
philosophy and science are all represented
in Canto's specially selected list of titles,
which now offers some of the best and
most accessible of Cambridge publishing to
a wider readership.

SUSAN ALDRIDGE

The Thread of Life

The story of genes and genetic engineering

CAMBRIDGE
UNIVERSITY PRESS

Published by the Press Syndicate of the University of Cambridge
The Pitt Building, Trumpington Street, Cambridge CB2 1RP
40 West 20th Street, New York, NY 10011–4211, USA
10 Stamford Road, Oakleigh, Melbourne 3166, Australia

First published 1996
Canto edition 1998

Printed in Great Britain at the University Press, Cambridge

A catalogue record for this book is available from the British Library

Library of Congress cataloguing in publication data
Aldridge, Susan.
The thread of life: the story of genes and genetic engineering / Susan Aldridge.
p.   cm.
Includes bibliographical references and index.
ISBN 0 521 46542 7
1. Genetic engineering.   2. Genetics.   3. DNA.   4. Biotechnology.
I. Title
QH442.A43   1996
575.1*0724—dc20   95-7354 CIP

ISBN 0 521 46542 7 hardback
ISBN 0 521 62509 2 paperback

Contents

Preface

The human genome project, DNA testing, gene therapy, and genetic engineering . . . there is no shortage of news about the gene revolution. This book aims to take you behind the headlines and explore the fast-moving and fascinating world of molecular biology.

In the first part of the book, I have tried to convey the power and uniqueness of the DNA molecule: how it was discovered, what it does, and where it came from. This leads into genetic engineering and its potential. In a very real sense, there is nothing special about gene transfer – it has been going on for billions of years. Its potential comes from humans, rather than the blind forces of evolution, being at the controls. The applications of genetic engineering and related technology that have attracted the most publicity – gene testing and therapy, and transgenic animals – are considered next.

But genetic engineering is just one aspect of biotechnology (although the two terms are often used synonymously); in the third part of the book I try to look at the wider world of biotechnology, as well as that of genetic engineering – as applied to plants and the environment.

Critics say that there is an overemphasis on DNA in biology, leading to a kind of reductionism, which has alienated the public, and some scientists, from its benefits. In the last part, I have tried to put DNA in context by looking at some other new ideas that have emerged in biology over the last 20 years or so.

Susan Aldridge
London

Acknowledgements

I am grateful to the following people for the help they have given me with this book: Professor Lynn Margulis of the University of Massachusetts and Dr Iain Cubitt of Axis Genetics for taking the time to talk to me about their work; Professor Ted Tuddenham of the Medical Research Council's Clinical Sciences Centre for providing me with an opportunity to learn about molecular genetics, and all my colleagues and friends in the lab at the CSC for five years of interesting discussions; Sandy Smith and my husband, Graham Aldridge, for reading the manuscript and making many helpful suggestions. Finally, thanks are due to Dr Robert Harington, formerly of Cambridge University Press, for his enthusiastic support at the outset of the project and Dr Tim Benton for his continuing help and guidance.

Illustration credits

Figs. 1.2(a) and (b) and 1.3(a) and (b) are modified from *Biological Science 1&2*, 2nd edition, by N. P. O. Green, G. W. Stout and D. J. Taylor, edited by R. Soper. Cambridge University Press, 1993.

Figs. 2.1, 2.2, 2.3, 4.1(a) and (b), 5.1, 7.1 and 10.1 are modified from *Biochemistry for Advanced Biology*, by Susan Aldridge. Cambridge University Press, 1994.

Fig. 3.1 is from *Gray's Anatomy*, 35th edition, edited by R. Warwick and P. Williams. Longman (London), 1973.

PART I

What is DNA?

1

DNA is life's blueprint

Take a large onion and chop finely. Place the pieces in a medium-sized casserole dish. Now mix ten tablespoons of washing-up liquid with a tablespoon of salt, and make up to two pints with water. Add about a quarter of this mixture to the onion and cook in a bain-marie in a very cool oven for five minutes, stirring frequently, and liquidise at high speed for just five seconds.

Now strain the mixture and add a few drops of fresh pineapple juice to the strained liquid, mixing well. Pour into a long chilled glass and finish off by dribbling ice-cold alcohol (vodka will do) down the side so that it floats on top of the mixture. Wait a few minutes and watch cloudiness form where the two layers meet. Now lower a swizzle stick into the cocktail and carefully hook up the cloudy material. It should collapse into a web of fibres that you can pull out of the glass. This is DNA (short for deoxyribonucleic acid).

DNA is the stuff that genes are made of. Genes carry biological information, which is translated into the characteristics of living things and is passed on down the generations. So genes determine the colour of a butterfly's wings, the scent of a rose, and the sex of a baby. DNA is just a chemical – not a more complex entity like a chromosome or a cell – and it is only in a biological context that it acquires its status as the molecular signature of an organism.

Discovering DNA

The chopping, cooking, grinding and mixing processes described above resemble those carried out every day in laboratories all around the world to extract DNA from living tissue. DNA dominates modern biology – yet many decades passed between its discovery and the realisation of its significance.

DNA was first isolated from human pus – a mixture of bacteria, blood plasma and white blood cells that exudes from infected wounds and abscesses – by the Swiss biochemist Friedrich Miescher in 1869. Miescher's life and work were greatly influenced by his uncle, the anatomist Wilhelm His, who was one of the founding fathers of molecular biology. After studying medicine, Miescher moved to Tübingen to work with the great chemist Felix Hoppe-Seyler in the first laboratory in the world devoted exclusively to the study of biochemistry.

The late nineteenth century was an exciting time. Although English physicist Robert Hooke had described cells as long ago as 1665 in his classic work *Micrographia*, it was only in the nineteenth century that their significance as the fundamental building blocks of all organisms was realised. Cells are tiny compartments, full of a fluid called cytoplasm and separated from their outer environment by a thin membrane of a fatty material. Organisms may consist of a single cell – like bacteria, amoebae and yeast – or they may exist as a community of different types of cell working together. These multicellular organisms range from sponges, jellyfish and tiny pond animals, which get by with just a few cell types, to humans who boast over 200 different types of cell.

In the 1860s the idea that life somehow arose spontaneously was finally overturned. Rudolf Virchow, the father of clinical pathology, developed the idea that cells – life's building blocks – could only come from other cells. Experiments carried out by the great French scientist Louis Pasteur supported this view. Pasteur showed that vessels containing broth went mouldy only when contaminated by airborne microbes. If they were heated and sealed they remained sterile – no microbial life appeared spontaneously under such conditions.

Bacteria come from other bacteria, by a simple cell division process called binary fission. This might happen as often as every 20 minutes. Given unlimited food and energy – and an idealised predator-free environment – a single bacterium would give rise to numbers greater than the human population (5 billion) in under 11 hours.

Binary fission is an example of asexual reproduction, in which a new organism comes from a single 'parent'. More complex creatures, such as ourselves, come from the union of a cell from each of two parents. This is sexual reproduction. The cell formed by this union grows into a complete organism (containing, in a human, at least $1\,000\,000\,000\,000$ or 10^{12} cells) by a cell division process called mitosis. In multicellular organisms, cells multiply rapidly only during development and in response to tissue damage. The rest of the time there is a balance between cell death and cell renewal.

Each time a cell divides it produces two cells of the same type. A human skin cell has to make another human skin cell, for example, and leaf cells must make more leaf cells, while bacteria produce more bacteria of the same species. The problem facing Virchow, Pasteur and their contemporaries was how to build on the proof that cells come from other cells and show how the particular characteristics of each cell type were transmitted when the cells multiplied.

Most cells are too small to be seen with the naked eye, so much laboratory time was spent peering at them through microscopes. The new biochemists seized upon the intensely coloured dyes, such as Perkin's Mauve, which were being turned out by the fledgling German chemical industry. These stains helped to reveal the inner structure of cells. This, together with improvements in the optics of the microscope, showed that many cells have a central core known as the nucleus (first observed in 1831). Just prior to Miescher's discovery of DNA, the German scientist Ernst Haeckel had suggested that the nucleus was of key importance in passing on characteristics from one generation to the next.

Miescher had a particular interest in the chemical contents of cells. Every morning he would call at the local clinic to pick up used bandages. In the days before antiseptics, these would be soaked in pus and Miescher had discovered that the large nuclei of the white blood cells it contained were ideal for his studies.

It was in these nuclei that Miescher discovered a new substance in 1869. It appeared only when he added an alkaline solution to his cells. Looking under the microscope he saw that this treatment made the nuclei burst open, releasing their contents. So he called the new substance nuclein, on the assumption that it came from the nucleus.

Analysis of nuclein showed that it was an acid, and that it contained phosphorus. These findings suggested that nuclein did not fit into any of the known groups of chemicals that are found in cells, such as proteins, carbohydrates and lipids. Miescher went on to show that nuclein was present in many other cells. Later, nuclein was renamed nucleic acid, and we now know it as DNA.

Miescher became particularly interested in sperm cells from Rhine salmon, because nuclei account for more than 90% of the cell mass in that type of cell (in later life Miescher's attention shifted back to the whole organism and he became interested in the conservation of Rhine salmon). In these experiments, Miescher also extracted a simple protein, protamine, from the nucleus. Protamine is unique to sperm nuclei. In all other nuclei, a similar protein is found called histone, first identified by the German chemist, Albrecht Kossel. It was therefore established that the nucleus contained both DNA and protein – but which of them was involved in the process of inheritance?

Picking up the threads of inheritance

Meanwhile, the microscope was revealing more and more about the secret life of cells. In 1879, the German chemist Walther Flemming discovered tiny thread-like structures within the nucleus, made of a material that he called chromatin because it readily absorbed colour from the dyes used to stain cells and tissues. (Later these threads were named chromosomes.)

The stained chromosomes revealed the intimate details of mitosis to Flemming and others. They saw how the chromosomes double up, as if a copy of each has been provided by the cell. And then, just before the cell divides, the paired chromosomes split up, like a divorcing couple, with each eventually taking up residence in one of the two new cells produced by the cell division. So mitosis is accompanied by the delivery of a fresh set of chromosomes to each new cell.

Then sex was put under the microscope. Oskar Hertwig, working in the French Riviera in 1875, placed tiny drops of sea water containing eggs and sperm from the Mediterranean sea horse on a glass slide, focussed the lens on his microscope, and then sat back to watch the action. He missed the moment of fertilisation – when sperm and egg cells meet – but saw their two nuclei fuse together and then begin to divide.

Eight years later Edouard van Beneden, at the University of Liège, saw that chromosomes from sperm and egg mingled during the fertilisation of the horse threadworm. More importantly, he saw that these germ cells had half the number of chromosomes that other cells had. As we now know, germ cells are formed from a special kind of cell division called meiosis, which involves a halving of the number of chromosomes. So when the germ cells join in fertilisation, the fertilised egg – which goes on to form a new organism – has a full complement of chromosomes.

Chromatin was, evidently, the stuff of inheritance. Analysis showed that it contained nuclein, and in 1884 Hertwig stated that 'Nuclein is the substance that is responsible . . . for the transmission of hereditary characteristics', which is more or less in keeping with our current understanding of the role of DNA.

Ironically Miescher, who speculated long and hard on the biological role of nuclein, could never accept Hertwig's ideas. He did, however, believe that information could be handed on from one cell to the next as a chemical code, stored in large molecules such as proteins. In 1892 he wrote to his uncle, Wilhelm His, that repetition of chemical units in such large molecules could act as a language 'just as the words and concepts of all languages can find expression in the letters of the alphabet'.

He remained devoted to the study of nuclein and worked long hours at the low temperatures he believed gave the best results. He was right about this – time has shown that DNA is a fragile molecule, and this is why chilling is so important in the extraction of DNA from the onion, described at the beginning of this chapter (and why a bucket of ice is a vital accessory for all self-respecting molecular biologists). In the end his intense efforts took their toll and his health, always delicate, broke down, leading to his death at the age of 51.

Dissecting DNA

So, between them, Hertwig and Miescher were right about DNA. But it has taken nearly a century of chemistry to turn their ideas into the central concept of molecular biology.

The chemistry of life is based on the element carbon. There are probably millions of different carbon compounds in nature. Most of them have never been identified and some – like the potential drugs made by plants on the verge of extinction – never will be. The chemical investigation of natural products, such as DNA, will never be over. Specialist chemical journals overflow with reports of exciting new compounds extracted from sponges sitting on the sea bed, common weeds, insects, and human tissue – to name but a few of their sources. Some of these compounds are interesting in their own right, because they have a novel arrangement of atoms, while others have an immediately obvious application to human health and welfare.

The goal of a chemist who extracts a natural product is to find out its structure; that is, the arrangements of atoms in the molecules of the new compound. Structure usually points to the properties and function of a substance. For example, the structure of diamond – which is a giant network of carbon atoms, each of which is linked to four others surrounding it – gives it the property of hardness. Thus, diamonds can be used to make hard-wearing cutting tools. The link between structure and function that was eventually found in DNA is without doubt the most significant in the history of chemistry.

The first step on the road to structure is to find out what elements the new compound contains and, from this, a crude chemical formula can be worked out. Miescher found a formula for DNA: $C_{29}H_{49}O_{22}N_9P_3$ (that is, for every 29 carbon atoms there are 49 hydrogen, 22 oxygen, 9 nitrogen and 3 phosphorus atoms).

The structures of simple molecules such as water are easy to solve, because a formula like H_2O (two hydrogen atoms and one oxygen) suggests only one possible arrangement for its atoms, according to the laws of chemistry. But Miescher's formula for DNA (which turned out to be incorrect – it was too simple) suggested thousands of possibilities. The next step, undertaken by Miescher and his contemporaries, was to break up the DNA molecule and find out what distinct groupings of atoms it contained.

One of the things that make people think chemistry is both boring and confusing is a lack of appreciation of the rules that make sense out of the pages of equations and formulae. There are 92 elements that occur in nature, but they do not just combine together at random to make billions of unrelated compounds. The rules about chemical combination, which were worked out in the nineteenth century, have led to a huge database of chemical families, each containing distinct groups of atoms. Compounds in the same family tend to behave in a similar, predictable way. For example, chlorides, such as sodium chloride and magnesium chloride, will always give a tell-tale white precipitate if they are mixed with silver nitrate.

Once the formula of a compound is known, the next step is to assign it to a family by doing a series of tests on it (like the silver nitrate test, described above). Complex substances such as DNA tend to have the attributes of more than one chemical family, because they contain more than one distinct group of atoms.

By 1900, chemists had worked out that DNA contained components from three different families of chemicals. It contained phosphate, a sugar – and a 'base'.

A phosphate group consists of a phosphorus atom surrounded by four oxygen atoms. The place you are most likely to see phosphate mentioned these days is in the supermarket, where rows of 'green' detergents boast of being 'phosphate-free'. Phosphate is an essential nutrient – just look on any packet of garden fertiliser – for all living things because it is an essential component of DNA (and of bone). If phosphates from detergents make their way into the water supply, they seem to encourage the luxuriant growth of algae in rivers, at the expense of other organisms, introducing an imbalance into the ecosystem. The reason why phosphates are present in detergents is that they stop dirt settling back onto fabric once it has been removed.

The sugars are a subgroup within a big biochemical family known as the carbohydrates. As the name suggests, they contain carbon and the elements of water – hydrogen and oxygen. It is easy to confirm this. Heat a teaspoon of table sugar over a flame and within minutes the white crystals give off clouds of steam, leaving a black puffy mass of carbon behind on the spoon. Or just watch toast burning – another example of the decomposition of carbohydrate! The 'D' in DNA stands for deoxyribose, the name of the sugar that was eventually identified as being part of the DNA molecule.

But it is the 'bases' that are the most significant part of the DNA molecule. Bases are chemicals that react with acids to neutralise them. One example of a base is ammonia, which is often an ingredient of household cleaners. Another is baking soda (sodium bicarbonate). If you add this to vinegar (an acid), neutralisation takes place, accompanied by effervescence, which is the evolution of carbon dioxide gas as a by-product. Something similar happens when you drop a tablet of the antacid Alka-Seltzer into a glass of water – the dry tablets contain citric acid and sodium bicarbonate; these react together when they come into contact with water.

The bases in DNA are a bit more complicated than those mentioned above, but, as they are so important in the storage and transmission of biological information, it is worth looking at them in some detail. They belong to two families called the purines and the pyrimidines. The purines in DNA are adenine (A) and guanine (G), while the pyrimidines are cytosine (C) and thymine (T). Like many biologically active molecules, such as vitamins and barbiturates (sedative drugs), the purines and pyrimidines contain carbon and nitrogen atoms arranged in a ring. Purines contain a hexagonal and a pentagonal ring, fused together, while the pyrimidines have just a single hexagonal ring. These family loyalties of the four bases in DNA – and, in particular, the relationships between the two families – were to be of immense importance in working out the detailed structure of the molecule. Put simply, purines and pyrimidines link up in DNA, providing the linchpin of the molecule and the key to its biological significance.

This, however, is getting ahead of the story: around the turn of the century, all the chemists had was the pieces of the DNA jigsaw – phosphate, deoxyribose and the bases. They then had to work out how these were linked together.

Much of the credit for this has to go to a brilliant and productive biochemist called Phoebus Levene. He studied under the chemist and musician Alexander Borodin at the Chemical Institute in St Petersburg before emigrating to New York with his family in 1891. There he settled at the new Rockefeller Institute for Medical Research. Levene, who was continually in trouble with the Rockefeller's director, Simon Flexner, for overspending his budget, set to work to analyse DNA more closely.

He showed that the three components were linked together by chemical bonds in the order phosphate–sugar–base, with the sugar forming a kind of bridge between the phosphate and the base. He called this unit a nucleotide, and argued that DNA was made of several nucleotides strung together like beads on a necklace. He went on to prove that the string of chemical bonds that linked the nucleotides together – the thread of the necklace – ran through the phosphate groups, not the bases.

By this time another kind of nucleic acid had been identified – in the cytoplasm. This is ribonucleic acid or RNA. RNA is similar in its chemical makeup to DNA, but a base from the pyrimidine family called uracil (U) is found in place of T, while, as the name suggests, the sugar ribose replaces deoxyribose.

DNA or protein?

Unfortunately, what was not appreciated until well into the twentieth century was the sheer length of the DNA molecule. If you were to extract the DNA from the chromosomes of a single human cell and piece it together as one molecule it would have a length of over two metres. In one of the simplest organisms, the bacterium *Escherichia coli (E. coli)*, the DNA molecule is just over a millimetre long – a thousand times longer than the diameter of the bacterial cell itself. DNA molecules are different lengths in different species – but even the smallest and shortest consist of thousands of nucleotides strung together.

Levene and his contemporaries pictured a much smaller molecule, with fewer than ten nucleotides. What probably led them to this conclusion was the DNA breaking up into small fragments during their experiments. A DNA molecule must be extraordinarily thin – being thousands, millions, or even billions of times longer than it is wide. So mechanical handling could easily snap such fragile fibres. In our rather low-tech DNA extraction from an onion, we try to preserve the integrity of the fibres by minimising the time they spend in the liquidiser (this step is essential to break the membranes around the nuclei and cells of the onion tissue to release the DNA).

It was not until the 1930s that two Swedish scientists, Torbjörn Caspersson and Einar Hammersten, used new methods for measuring the size of the DNA molecule and showed it was a polymer. A polymer is a long molecule made up of a number of smaller molecules called monomers, chemically linked together. In DNA the monomers are nucleotides of four kinds, each containing a different base. In some polymers there is only one type of monomer. For instance, polythene is based on the monomer, ethene, a small hydrocarbon. The world's most abundant natural polymer, cellulose, is based on glucose as a monomer.

If genetic information is somehow encoded in the arrangements of atoms in a molecule, then that molecule is likely to be a polymer – given the amount of data that must be needed to specify the characteristics of even a simple organism. But the biological significance of DNA continued to be overlooked. Progress was held back by Levene's insistence that the amounts of the four bases were the same in all DNA molecules. He imagined a regular arrangement of the four corresponding nucleotides, each with its own base, along the whole molecule. The sequence Levene imagined for DNA went something like this – ACGTACGT-ACGT – endlessly repeating itself. Such an arrangement had no informational capacity (whereas a sequence with more variety, such as CCTATTTGAGTAA, would have). Levene's belief acquired the status of dogma and was known as the Tetranucleotide Hypothesis. It led to the assumption that DNA played a (literally) supporting role in heredity and merely held the all-important nuclear proteins in position.

As biochemical techniques developed during the early years of the twentieth century, proteins began to compare favourably with nucleic acids as candidates for life's blueprint. Where nucleic acids seemed simple, proteins – which are also polymers – were complex, subtle molecules. Levene's contemporary, the great German chemist Emil Fischer, showed that the monomers of proteins were simple molecules called amino acids. There are 20 amino acids commonly found in proteins, so when these are strung together in their hundreds – or even thousands – to make a protein molecule, there are an enormous number of possible sequences. Compare this with the monotonous regularity of Levene's DNA, and it is easy to see how the biochemists allowed themselves to be distracted.

Of course proteins are enormously important to cells. Every living cell is a hive of chemical activity where thousands of different molecules are either being broken down into simpler components, or built up into something more complex. Indeed while Levene and his contemporaries were speculating on the problems of inheritance, other biochemists were pursuing the equally thorny question of how on earth all this frantic cellular biochemistry was controlled and co-ordinated.

A typical bacterial cell might contain around 3000 different proteins, while a human cell could possess between 50000 and 100000. Of these perhaps half are biological catalysts called enzymes. They do two things: they all speed up chemical reactions, and each enzyme generally acts on one reaction only.

The human digestive system is a set of organs – stomach, pancreas and so on – whose cells are equipped with a set of digestive enzymes for breaking down food-stuffs and extracting their energy and nutrients. Without enzymes you would soon starve, because it would take 50 years to digest a typical meal. And you must have the full set of enzymes. Pepsin, for instance, which is found in the stomach, will only aid the chemical breakdown of protein to amino acids. It will act on the milk protein in a bowl of cornflakes, but will leave the breakdown of the cereal carbohydrates, and the fat in the milk, to other members of the digestive enzyme family.

In comparison to the digestive system, the chemical reactions typically occurring within the cell are far more complicated. In a liver cell there could be hundreds of reactions going on at any one time and each is catalysed by its own enzyme.

Soon enzymes were being given the status that rightfully belonged to DNA. The picture that emerged was of a nucleus containing a master set of enzymes held in position by a DNA support, from which working copies were generated by a direct copying mechanism. This 'Protein Dogma' was a powerful one and a few biochemists still clung to it as late as the mid-1950s before the role of DNA had been completely worked out.

Time for a paradigm shift

Ideas in science come in fashions called paradigms, where a paradigm is the current idea of how some aspect of the physical world works. When a scientific paradigm is first challenged, the scientific community generally responds with either furious denial or complete lack of interest, and the new hypothesis has to wait for many years to be 'rediscovered' when the time is right. (The theory of continental drift put forward by Alfred Wegener in 1912 is a good example. It was not until the 1960s that firm evidence of the movement of the earth's plates over geological time finally made Wegener's ideas acceptable.)

In 1944 Oswald Avery and his team (Colin McLeod and Maclyn Macarty) announced the discovery that DNA, not protein, carried life's blueprint. Avery, a medical microbiologist at the Rockefeller Institute, was a modest man, unlikely to make the flamboyant claim that he had discovered the secret of life. Yet his discovery was greeted by a standing ovation from his colleagues. This tribute must have been especially gratifying for Avery, coming as it did on the eve of his retirement. The years have enhanced the stature of his work, which would almost certainly have earned him a Nobel Prize had he lived longer.

Outside the Rockefeller, the response was muted. There was even a hint of pique that mere medical men, rather than pure scientists, had finally discovered the key role of DNA. However, few doubted the importance of Avery's results, and the task of unravelling the secret of how DNA worked jumped to the top of the scientific agenda.

What started Avery on the road to DNA was his determination to solve a problem posed in 1928 by Fred Griffith, a public health microbiologist working in London. Griffith had produced a start-ling result from his work with pneumococci, the bacteria which cause pneumonia. This disease was much feared in those days, before antibiotics, because it was so often fatal. Griffith showed that there were two different forms of pneumococcus, which he called rough and smooth from the appearance of the colonies each produced when growing on nutrient plates (a colony is just a community of millions of bacteria that is visible to the naked eye). The smooth form has a slimy coating of carbohydrate on the surface of its cells. This seems to act as a disguise when the bacteria enter the body. The rough form, lacking this coat, is recognised and rapidly disposed of by the body's defences and so is harmless.

Like most bacteria, the smooth form can be killed by heat; but when Griffith tried mixing heat-killed smooth pneumococci with live but innocuous rough bacteria, he found that this cocktail was lethal to his experimental mice. When he examined the animals' blood he found it laden with the smooth bacteria. Not only had the dead bacterial cells somehow passed their virulence to the smooth pneumococci, but this characteristic must have been transmitted to the progeny of the live bacteria.

Griffith had just performed the first documented experiment in genetic engineering but he did not, of course, recognise it as such. Genetic engineering very rarely happens in this spontaneous fashion (as we will see in Chapter 5). Griffith was perplexed by his results. The scientific community named the substance that had passed from the dead to the live pneumococci the 'transforming principle'.

Whatever this transforming principle was, it contained the instructions for making the carbohydrate coat so that rough bacteria could become smooth – and lethal. Avery had no illusions about the difficulty of extracting it from the pneumococcus cells.

'The . . . extract . . . is full of polysaccharide, carbohydrate, nucleo-proteins, free nucleic acids, lipids and other cell constituents. Try and identify in this complex mixture the active principle!', Avery wrote to his brother, another medical microbiologist, in 1943.

It took Avery and his team years of painstaking work to show that DNA was the transforming principle. The main plank of their project was the use of enzymes. The enzymes acted as a bio-chemical tool kit that could chop up specific cell components, leaving others unaffected. When the researchers added enzymes that digested proteins, the transforming activity was unaffected. So the transforming principle was not a protein. This was the first surprise. Then they added, in turn, enzymes that broke up carbohydrate, lipid and RNA. None of these could destroy the transforming principle. It was only when they added an enzyme that chops up DNA that the transforming activity finally dis-appeared.

In the excitement following the announcement of these results, many scientists rushed to their benches to try to verify Avery's results. This flurry produced another landmark experiment. In the 1940s it had become fashionable to work with bacteriophage (phage, for short) to study the chemical basis of genetics. Phage are viruses that attack bacteria, and many molecular biologists of this era flocked to a famous summer school on phage that was held at the Cold Spring Harbor, Laboratory in Long Island, New York.

Alfred Hershey, a member of the so-called 'phage group' at Cold Spring Harbor, said, in a letter to a colleague in 1951, 'I've been thinking . . . that the virus may act as a little hypodermic needle full of transforming principle.' Like any virus, when phage invade cells they reproduce themselves with the aid of the cell's enzymes. Hershey, with his colleague Martha Chase, proposed to let phage infect bacteria and identify the blueprint by finding out whether DNA, or protein, had entered the cell.

In what has now become their famous experiment (Fig. 1.1), Hershey and Chase exploited the chemical differences between DNA and protein. Only DNA contains phosphorus; only proteins contain sulphur. By growing phage in a chemical solution contain-ing radioactive phosphorus and radioactive sulphur, Hershey and

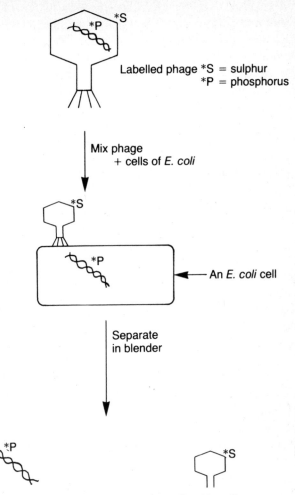

Labelled phage *S = sulphur
*P = phosphorus

Mix phage
+ cells of *E. coli*

An *E. coli* cell

Separate
in blender

Inside cells

Outside cells

Fig. 1.1. Hershey and Chase's experiment. Genetic material from bacteriophage, which contains the instructions for making more phage, must pass into bacterial cells during infection. Hershey and Chase's experiment shows that this material is DNA not protein because, in their doubly 'labelled' phage, the DNA contains radioactive phosphorus and the protein radioactive sulphur. After infection, radioactive phosphorus, rather than sulphur, leaves its tell-tale trace within the bacteria.

Chase ensured that the DNA and protein in the phage both acquired a radioactive tag. The presence of the tag meant that it would be easy to trace the fates of the phage DNA and protein after infection.

They mixed the phage – wearing its two radioactive tags – with the bacteria and allowed some time for the infection to occur. This left 'empty' phage attached to the outside of the bacterial cells, while the phage contents, with their genetic instructions for making more phage like themselves, ended up inside the bacteria. Hershey and Chase then tipped the mixture into a Waring blender – a kitchen food blender that was ideal not just for making soup but for severing the connection between the bacteria and the empty phage. They then broke open the bacteria to see what was inside and found the tell-tale radioactive signature of phosphorus – not sulphur. The phage had, therefore, injected DNA into the bacteria. The radioactive sulphur turned up in the empty phage, because the protein had merely provided a protective coat for the DNA – it had no instructions for the phage to reproduce itself.

Hershey and Chase had underlined Avery's discovery of DNA as life's blueprint. These discoveries caused a shift in researchers' opinions – in favour of DNA being the molecule transmitting information between generations. This paradigm shift now generated a new set of problems for those who accepted it – in particular, how did DNA carry information, and how was it translated into the biochemical activity of the cell? These two questions were central to biology for the next 20 years, and even today no molecular biologist would claim to have complete answers to them.

Deeper into DNA

It was the Austrian chemist Erwin Chargaff who made one of the major contributions to understanding DNA. Avery's work had made a deep impression on him. He wrote, 'I saw before me in dark contours the beginning of a grammar of biology. Avery gave us the first text of a new language, or rather he showed us where to look for it. I resolved to search for this text.'

By the 1940s analytical techniques were sufficiently advanced to allow the new insights into the structure of DNA that Chargaff was seeking. One of these – paper chromatography – was to prove particularly powerful in his search. Nowadays the paper chromatography of biological molecules such as chlorophyll and amino acids is part of school science. You could even do it at home, armed with only simple equipment.

Paper chromatography (the word chromatography means 'writing with colour') depends on applying a sample of a mixture of molecules – such as the pigments extracted from flower petals – to a piece of filter paper. The molecules attach themselves to the cellulose in the paper, but with differing affinities depending on their chemical nature. Some cling on fast to the cellulose; others are only weakly held. If one end of the filter paper is then dipped into a vessel containing a solvent, such as water or alcohol, the solvent will travel up the paper through the channels between the fibres of cellulose. As the solvent moves up the paper – and this movement can be seen quite clearly – the components of the mixture are swept along with it. Those that are weakly attached to the cellulose travel further up the paper than those that bind more strongly. The result is the spreading out of the components on the paper, depending on their affinity for cellulose.

Once the solvent has travelled up the paper, it is dried and the positions of the components on the paper are located with a visualising agent – such as ultraviolet light, which gives a fluorescent glow with most biological molecules. Sometimes the components are coloured. Chromatography of an extract from spinach leaves, for example, will give spots or bands of green, grey and orange on the filter paper, showing that green leaves contain a number of different pigments. So chromatography allows the separation of mixtures of chemicals. Further chemical detective work will lead to the identification of each component in the mixture.

Chargaff pioneered the paper chromatography of nucleic acids. He chopped DNA samples into their component nucleotides by incubating them with enzymes. He then subjected the resulting mixture to paper chromatography, getting four distinct spots or bands on the paper after visualisation, one for each nucleotide (remember that the nucleotides of DNA differ just by the base they each contain – A, C, G or T). Then he cut each band out of the filter paper and washed the nucleotides off with solvent, so that he could measure how much of each one there was in the DNA sample.

His work demolished Levene's tetranucleotide hypothesis. There were not equal amounts of each nucleotide in DNA after all. Analysis of DNA from yeast, bacteria, ox, sheep, pig and humans showed that each species could, instead, be characterised by the relative amounts of A, C, G and T in its DNA (this is the same as the relative amounts of the corresponding nucleotides). For example, in human DNA, 30.9% of the base content is A, whereas in yeast 27.3% of the base content is A. Within an individual species, these figures are identical, whatever tissue the DNA is extracted from. So if the DNA from the bulb of an onion – whose extraction was described at the beginning of this chapter – had been analysed, its base content profile would have been the same if the experiment had been repeated using onion leaves. This was interesting, and just what you would expect of a chemical blueprint – different between species, but the same within an individual species.

But Chargaff's main contribution turned out to be the discovery that the number of molecules of A in *any* DNA sample was always equal to the number of molecules of T – and likewise, the molecular amounts of C and G were also equal to one another. When this work was published in 1950, Chargaff seemed to make little of this finding – known later as Chargaff's ratios. It was left to others to seize upon the significance of his work.

The double helix

The double helix has become the symbol of molecular biology: from textbook covers and conference logos, to T-shirts and mugs given out by biotechnology companies, the DNA spiral is a constant reminder, to those who work in the field, of the powerful link between chemical structure and function.

All too often today, chemistry has a poor public image. Besides being blamed for environmental pollution (it was a chemist, Thomas Midgeley, who invented both leaded petrol and ozone-destroying chlorofluorocarbons or CFCs), chemists are generally seen as dull, analytical and lacking in imagination. No one seems to want to talk to a chemist at parties – except another chemist. In fact, the appreciation of chemistry requires an imaginative willingness to enter and explore the molecular world. It took a marriage between this kind of vision, and the right technology, to breathe life into the structure of DNA.

The technique of X-ray crystallography allows the creation of three-dimensional images of large biological molecules such as proteins and nucleic acids in which the position of every atom is fixed in space. Such information is not available from traditional chemistry-based techniques, which give merely a two-dimensional picture of a molecule with few clues as to how that molecule actually works.

Pioneered by the father and son team William and Lawrence Bragg in 1913, X-ray crystallography depends on the interaction of X-rays with the electrons surrounding the atoms in a crystal. A beam of X-rays is directed onto a crystal and, when it hits its target, this interaction causes it to deflect from its original direction. If the crystal is surrounded by photographic paper (which responds to the presence of X-rays), a pattern of dots due to the deflected X-rays is picked up on the paper. Mathematical analysis of this pattern is used to work out the three-dimensional arrangement of atoms in the crystal, giving a good impression of the overall shape of the molecule.

The Braggs began by looking at the regular arrangements of atoms in simple substances such as common salt (sodium chloride). Then other scientists, such as Max Perutz and Dorothy Hodgkin working in Cambridge, developed the techniques for a range of biologically active molecules including haemoglobin, vitamin B_{12}, and penicillin. By 1938, William Astbury, a student of the elder Bragg, had produced the first X-ray photograph of DNA. This proved hard to interpret, and it was not until the late 1940s that three separate groups began to work in earnest on the problem of the DNA structure.

At King's College in London, Maurice Wilkins (who, like many physicists, had turned to biology after working on the Manhattan Project, building an atomic bomb, during the Second World War) wondered if the long fibres that DNA forms when it is pulled out of watery solutions hinted at some regularity in its molecular structure. Using a makeshift X-ray apparatus, he produced new pictures – far superior to Astbury's. But again, the pictures were hard to interpret.

In 1951 Wilkins was joined by Rosalind Franklin, an expert in X-ray crystallography. Franklin built a new X-ray laboratory at King's and took some superb pictures of DNA. Quite early on she had the notion that the molecule was probably coiled into a helical shape. The helix as a motif in biological molecules was first put forward by the American chemist Linus Pauling, who had laid down the theory of chemical bonding. So the idea that DNA might also be helical was not really startling.

Meanwhile Francis Crick and James Watson had joined forces at the Cavendish Laboratory in Cambridge – united by little but their interest in DNA and a lack of knowledge of how chemical bonding worked. Each man had an impressive background, however – Crick in maths and physics, while Watson was one of the rising stars in the world of molecular biology. They set about the task of building a physical model of the DNA molecule, using metal plates and rods to simulate the nucleotide units and the chemical bonds between them. This partnership was housed in prefabricated huts beside the Cavendish, which are now used as bike sheds.

Crick and Watson were pursuing the idea that DNA was, in some way, a self-replicating molecule that copied itself during cell division to pass genetic information into new cells. Somehow a helical structure had to include this property of the molecule. They began to wonder if the bases were responsible for this unique feature of the molecule, perhaps by pairing up in some way. When Chargaff visited the Cavendish in 1952 and reminded Crick of his 1950 paper in which the base ratios were described, the fuse was lit that led to the final solution. Crick realised that A must pair with T, and C with G.

It was the news of a third entry into the DNA 'race' that accelerated Crick and Watson's efforts. When they heard that

Linus Pauling had just published his own DNA model – and that it contained a major error – the two men were determined to produce the correct model without delay. Wilkins was friendly with the pair, so Watson rushed down to London to show him Pauling's DNA paper.

Wilkins confided that relations between himself and Franklin were strained and that they were making little progress on the DNA project. He then showed Watson one of Franklin's X-ray pictures and, in Watson's words, 'The instant I saw the picture, my mouth fell open and my heart began to race.' The X-ray pattern seemed to point without doubt to a helical structure.

Back in Cambridge, the model building started in earnest. This time Crick and Watson decided to try a double helix to incorporate their ideas about base pairing. But still something was missing. They did not know whether the phosphates of the nucleotides should be inside or outside the double helix. Though they were convinced that the bases paired up, they could not see what could hold them together.

It was another visitor to the laboratory, the chemist Jerry Donohue, who solved the dilemma by pointing out that weak chemical links called hydrogen bonds could form between A and T, and between G and C. Hydrogen bonds are found in carbo-hydrates and proteins too, and between water molecules.

The hydrogen bonds between A and T, and G and C keep the purine-pyrimidine pairs together. Here was the underlying reason for Chargaff's ratios: wherever there is an A, there must be a corresponding T – because they are paired up in the DNA double helix – and the same is true for C and G (Fig. 1.2a). This was the final, and most important, piece of the DNA puzzle. Crick and Watson rapidly built the definitive model of DNA. It was a double helix, with the bases on the inside, facing one another in their pairs, while the phosphates ran down the outside, forming the backbone of the molecule (Fig. 1.2b). Deoxyribose was the bridge that held the phosphate and base together. A rough analogy would be a spiral staircase where the phosphates are the rails and the base pairs the steps.

The beauty of this model is that the structure immediately suggests function. As Crick and Watson hinted, in their

(a)

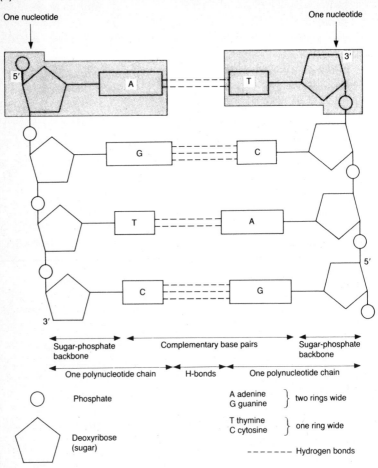

Fig. 1.2. (*a*) DNA diagrammatic structure of straightened chains. Each nucleotide unit in the DNA molecule is made up of three parts: phosphate, sugar and base. The bases pair together by hydrogen bonding down the centre of the two chains – A with T and C with G. (*b*) The DNA double helix. The DNA molecule consists of two long chains of nucleotides twisted around one another to make a double helix. The dimensions of this helix are shown (1 nm is 0.000000001 metre).

(*b*)

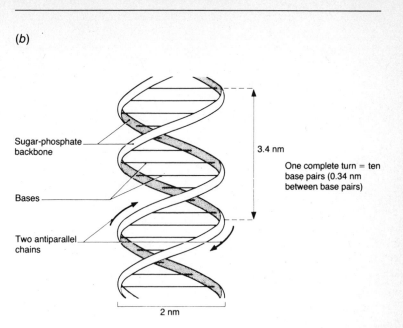

publication of the double helix structure, in the journal *Nature*, on 25 April 1953, 'It has not escaped our notice that the specific pairing we have postulated suggests a possible copying mechanism for the genetic material' (Fig. 1.3*a*).

In 1957 this suggestion was followed up in an elegant experiment by Matthew Meselson and Franklin Stahl, working at Woods Hole Marine Biological Laboratories in Massachusetts (Fig. 1.3*b*). They fed a colony of *E. coli* with a nutrient containing a so-called heavy isotope of nitrogen, whose atoms weighed more than those of normal nitrogen. The bacteria consumed this nutrient, using it to build up DNA as they multiplied. So this new DNA had the heavy form of nitrogen incorporated into its bases. Meselson and Stahl extracted this DNA from the bacteria, using techniques similar to those described at the start of this chapter. Then a sample of bacteria from the colony growing on the nutrient that contained heavy nitrogen was removed and transferred to nutrient

(a)

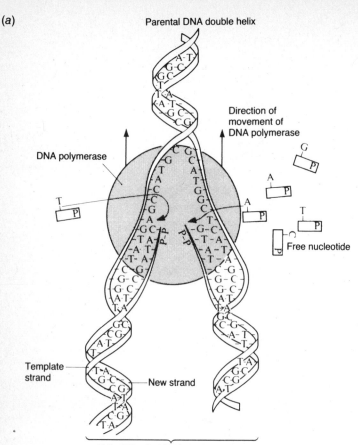

Fig. 1.3. (*a*) Simplified diagram of the replication of the DNA double helix. As the enzyme DNA polymerase 'unzips' the double helix, single nucleotides move into position next to their complementary bases. This creates two new strands, as shown at the bottom of the diagram. (*b*) Meselson and Stahl's experiment. The single nucleotides contain 'light' nitrogen so the two 'new' strands created in (*a*) contain 50% 'heavy' and 50% 'light' nitrogen. As these strands replicate in turn, the proportion of 'light' nitrogen in the DNA sample increases, as shown.

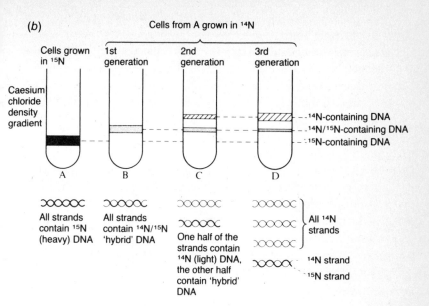

(b)

Cells from A grown in ¹⁴N

Cells grown in ¹⁵N | 1st generation | 2nd generation | 3rd generation

Caesium chloride density gradient

¹⁴N-containing DNA
¹⁴N/¹⁵N-containing DNA
¹⁵N-containing DNA

A   B   C   D

All strands contain ¹⁵N (heavy) DNA

All strands contain ¹⁴N/¹⁵N 'hybrid' DNA

One half of the strands contain ¹⁴N (light) DNA, the other half contain 'hybrid' DNA

All ¹⁴N strands

¹⁴N strand
¹⁵N strand

that contained the normal, light form. Again the bacteria multiplied and Meselson and Stahl extracted DNA. They did this until they had DNA samples from several generations of these bacteria.

Then each DNA sample was analysed to see how heavy it was. This was done by suspending the DNA extract in a salt solution and then whirling it in a centrifuge. The DNA molecule sank to a level in the tube that was indicative of its weight. Heavy DNA molecules tended to sink, light ones to float.

The first generation of DNA was 50% light and 50% heavy. In the next generation, the corresponding weight distribution was 75% light, and 25% heavy. Meselson and Stahl argued that this pointed to so-called semi-conservative replication.

The original DNA contains only heavy nitrogen. Just before replication, it splits into two single strands. These immediately attract the raw materials of DNA, the A, T, C and G-containing nucleotides that are present in all cells. These nucleotides – containing light nitrogen, because this occurs after the cells have been transferred from the heavy to the light nitrogen nutrient – line up in position along each strand. An enzyme clips them together through their phosphate groups. Each new double helix now has one old and one new DNA strand, hence the term semi-conservative replication. Each contains 50% heavy and 50% light nitrogen, just as Meselson and Stahl observed. In the next generation there would be four single DNA strands to be copied. The four new double helices created after replication would now contain two single strands with heavy nitrogen and six with light nitrogen – again corresponding with Medelson and Stahl's 25% and 75% ratios for the second generation.

This copying relies on base pairing through hydrogen bonding, and base pairing has turned out to be the most powerful concept in molecular biology. Not only is it at the heart of DNA replication – and the passage of genetic information through nearly four billion years of evolution – it also drives gene expression (the way DNA makes proteins), genetic engineering and all the other DNA-based technologies.

2
DNA in action

DNA is a database. The information it contains allows the assembly of the all-important protein molecules within the living cell from their component parts, the amino acids. This database is passed on from one cell to another by the DNA molecule's power of self-replication, as we saw in Chapter 1. Now we look at how the protein recipes are extracted from DNA during the lifetime of an individual cell.

Information flow in most organisms is a one-way street: from DNA to protein. This rule is known as the Central Dogma of molecular biology. This term was coined by Francis Crick in 1956, many years before the details of the molecular processes involved were actually worked out.

The genetic code

The concept of the gene as an inheritable factor responsible for an organism's characteristics was first put forward by the Austrian botanist and monk Gregor Mendel in 1865. Mendel looked at how characteristics such as flower colour, height and seed shape were inherited during carefully controlled experiments with pea plants.

Peas are normally pure-breeding. They reproduce by self-pollination, in which pollen and ovules (the male and female sex cells) from the same plant unite, to give offspring that are similar to their parent. So a pea with white flowers would normally produce more peas with white flowers, and so on.

Mendel removed the stamens (which produce pollen) from his

experimental plants, so they could no longer self-pollinate. Then he chose pairs with different characteristics: smooth or wrinkled seeds, red or white flowers, for example. He cross-pollinated these pairs by brushing pollen from one plant onto the stigma (the tip of the female sex organ) of the other. Then he planted the seeds that developed, and waited to see what types of plant his experiments had produced.

Common sense might suggest that this new generation would blend the characteristics of their parents – offspring of plants with wrinkled and smooth seeds should have seeds that are slightly wrinkled, for example. Mendel found that, on the contrary, there was no averaging out of characteristics. Either one characteristic or the other was inherited.

Mendel's careful experiments with thousands of pea plants led to the observation of distinct patterns of inheritance, which is discussed in Chapter 4. For now, the important feature of Mendel's work is his suggestion that physical characteristics are associated with 'factors' that are carried down the generations. Today we call these factors 'genes' and we know they are located in the chromosomes of the nucleus. We can pin it down even more precisely than this: a gene is a segment of DNA.

Mendel, however, had no idea of the physical nature of his factors. Perhaps this is why his work created little impact. He knew Darwin's work on evolution (which is discussed in Chapter 4) but it seems likely that Darwin was not aware of Mendel's achievements. It was probably religion that prevented Mendel and Darwin coming together to concoct some mechanism for evolution. The scientific community may have ignored Mendel's work, but it did not escape the eagle eye of the religious authorities. Fearful of reprisals that would threaten his position – he became abbot of his monastery in 1868 – Mendel made little effort to publicise his work. It was not until 1900 that several scientists repeated and confirmed his experiments, and his great contribution to the science of genetics was recognised.

During the first half of the twentieth century, efforts to pin down the nature of the gene gathered pace, starting from the general understanding that genes were located in the chromosomes. So while some biochemists – Levene, Avery, Hershey and Chase –

were busy trying to find out whether genes were made of DNA or protein, others were preoccupied with different questions, such as how large genes were, and whether they were actually molecules, or bigger entities such as the chromosomes themselves.

Inspiration was needed, and it came from the world of physics. Max Delbrück, working in Berlin, joined forces with a team of geneticists who were carrying out experiments in the Mendelian tradition on fruit flies (*Drosophila*). They were looking at mutations – changes in the chromosomes that led to observable changes in the characteristics of the flies' offspring. They bombarded the insects with X-rays to increase the mutation rate. From this work emerged an estimate of the size of the gene – a thousand atoms or fewer – which meant that genes had to be molecular in nature (although their estimate was too small; a typical gene contains around 1000 nucleotides and each nucleotide has around 50 atoms).

Delbrück borrowed ideas from the exciting new world of quantum physics (which would have been unfamiliar to most biologists) to explain why X-rays produce mutations. The gene was a stable molecule in its normal environment; but if it encountered an X-ray, which has much higher energy than its normal surroundings, it can undergo an irreversible change, or mutation, which changes the characteristics of the organism. So an X-ray impact on a *Drosophila* gene for eye colour might give a fly with white eyes, rather than the usual red eyes. We now know that this change occurs because the gene that creates the pigment in the flies' eyes is changed so it no longer functions. Thus, flies with this mutation have unpigmented (white) eyes rather than the usual pigmented (red) ones.

Delbrück was friendly with Nobel Prize winner Erwin Schrödinger, who was one of the founding fathers of quantum physics. Schrödinger had a long-standing interest in biology, and his discussions with Delbrück led him to set out his own thoughts on the nature of the gene in a lecture series in Dublin in 1943.

Schrödinger had moved to Dublin in 1938 to escape the Nazi invasion of his native Austria. He was a popular figure in Ireland because of his obvious enthusiasm for Irish culture, as well as his scientific stature. The lecture series was attended by Prime

Minister de Valera and drew a crowd of influential figures from the worlds of politics, the arts, and the Church. Later he published the lectures as a book entitled *What is Life?*

Schrödinger put forward the idea that the gene has a special molecular status. In his own words, 'the most essential part of a living cell – the chromosome fibre – may suitably be called an aperiodic crystal'. He went on to explain that the crystals that physicists and chemists were so used to dealing with were characterised by a regular inner arrangement of atoms or ions. In sodium chloride, for instance – which was investigated by the Braggs, as discussed in Chapter 1 – each sodium ion is surrounded by six chloride ions, and vice versa. Schrödinger called such an arrangement a periodic crystal. Its information content is extremely low, because there is no variety in the arrangement of the ions. By contrast, an aperiodic crystal has a high information content. Schrödinger put it like this 'The difference in structure is of the same kind as that between an ordinary wallpaper in which that same pattern is repeated again and again in regular periodicity and a masterpiece of embroidery, say a Raphael tapestry, which shows no dull repetition but an elaborate, coherent, meaningful design traced by the great master.'

Schrödinger suggested that a chemical code could be embedded in the gene. He saw the aperiodic crystal as a long, linear molecule made up of small units (which we now know to be nucleotides), which acted as the 'letters' of this chemical code. Following this idea through he said ' . . . with the molecular picture of the gene, it is no longer inconceivable that the miniature code should precisely correspond with a highly complicated and specified plan of development and should somehow contain the means to put it into operation'. James Watson commented on the book, 'From the moment I read Schrödinger's *What is Life?* I became polarized towards finding out the secret of the gene', and this perhaps sums up the enormous impact it had upon the emerging new generation of molecular biologists.

As for the chemical nature of the gene, however, Schrödinger had put his money on protein. So the experimental vindication of his ideas had to await the confirmation that genes were made of DNA.

Soon after Crick and Watson published their structure of DNA, physicist George Gamow suggested that the bases could work as a four-digit code. The order, or sequence, of the bases in an organism's DNA is, as Gamow put it, 'the signature of the beast' – which somehow encodes all its characteristics (remember that the units that actually make up DNA are nucleotides, but the bit of the nucleotide that actually varies is the base – so from now on we will talk about bases when discussing the genetic code).

By this time, it was becoming clear that the type of information encoded by DNA must be the acid sequences of proteins. But there are only four bases, while there are 20 amino acids commonly found in proteins. Obviously, the simplest code – a one to one correspondence (e.g. thymine equals glycine, the simplest of the amino acids) would not do, as this would leave 16 amino acids out in the cold. Gamow, who had published papers on the Big Bang theory as well as popular science books featuring a character called Mr Tompkins, began to speculate on how the bases could combine together in DNA to form a genetic code.

So one base to each amino acid will not work. What if the bases pair up, with each pair specifying one amino acid? This gives 4^2 or 16 different arrangements, insufficient to code for 20 amino acids. A triplet code gives 64 (4^3) possible arrangements, while a quartet code gives 256.

The problem of the genetic code was taken up by Crick and the South African molecular biologist Sydney Brenner. By 1961 they had experimental proof that the code is made up of non-overlapping triplets. Brenner coined the term 'codon' for a triplet of bases. In the same year Marshall Nirenberg and Johann Matthei, working at the National Institutes of Health in Washington, identified the first 'letter' in the code. In a test tube set-up that contained all the chemical components needed to make proteins from genes they identified the codon that coded for the amino acid phenylalanine. This codon had the sequence AAA – three adenines in a row. The rest of the code was soon broken by similar experiments.

Of the 64 possible codons (remembering that a triplet arrangement gives 4^3 possibilities according to the above arrangement), 61 code for the 20 amino acids. All of the amino acids, bar

Table 2.1. *The genetic code*

UUU	Phe	UCU	Ser	UAU	Tyr	UGU	Cys
UUC	Phe	UCC	Ser	UAC	Tyr	UGC	Cys
UUA	Leu	UCA	Ser	UAA	Stop	UGA	Stop
UUG	Leu	UCG	Ser	UAG	Stop	UGG	Trp
CUU	Leu	CCU	Pro	CAU	His	CGU	Arg
CUC	Leu	CCC	Pro	CAC	His	CGC	Arg
CUA	Leu	CCA	Pro	CAA	Gln	CGA	Arg
CUG	Leu	CCG	Pro	CAG	Gln	CGG	Arg
AUU	Ile	ACU	Thr	AAU	Asn	AGU	Ser
AUC	Ile	ACC	Thr	AAC	Asn	AGC	Ser
AUA	Ile	ACA	Thr	AAA	Lys	AGA	Arg
AUG	Met	ACG	Thr	AAG	Lys	AGG	Arg
GUU	Val	GCU	Ala	GAU	Asp	GGU	Gly
GUC	Val	GCC	Ala	GAC	Asp	GGC	Gly
GUA	Val	GCA	Ala	GAA	Glu	GGA	Gly
GUG	Val	GCG	Ala	GAC	Glu	GGG	Gly

These are the 64 codons and their corresponding amino acids and 'stop' messages. By convention, the genetic code is written in the language of messenger RNA (mRNA). So AAA on a gene will 'translate' to UUU in mRNA and this, eventually, will lead to the insertion of the amino acid Phe (phenylalanine) in the corresponding protein. Phe, Leu, etc. are the conventional three-letter abbreviations for the 20 amino acids that occur in proteins.

methionine, have more than one corresponding codon (Table 2.1). For example, both AAA and AAG code for phenylalanine, and some of the amino acids have four, or even six, codons. These multiple codons are said to be degenerate, but each codon is unambiguous – only one amino acid is specified by each one.

The three codons that do not code for amino acids are called stop codons. These indicate the end of the amino acid sequence which a particular DNA sequence codes for. Statistically, you would expect to come across one stop signal in every 20 codons (roughly 3 in 64). But most proteins have at least 100 amino acids – some have thousands. A stretch of DNA that codes for protein keeps going until the whole of the amino acid sequence has been specified. This kind of sequence is called an open reading frame (ORF).

As we shall see, not all DNA codes for protein. Scanning through a DNA sequence for ORFs – either by eye or by using a pattern-matching program on a computer – enables the coding sequences to be picked out. We can now formulate a definition of a gene in chemical terms, and pin down the more abstract concept put forward by Mendel and Schrödinger: a gene is a stretch of DNA that codes for a particular protein. For instance, the gene for the enzyme amylase, which breaks down starch molecules in bread when you chew a sandwich, is a stretch of DNA carrying the code for the amino acid sequence of this important molecule.

With a few exceptions, the genetic code is universal. Organisms as diverse as the bacterium *Escherichia coli*, higher plants and humans use the same DNA dictionary to translate the messages in their genes. This is one of the strongest proofs we have for the common ancestry of all life or monophyly. It makes sense that the genetic code had stayed the same for billions of years – a mutation that changed it would result in proteins with the wrong amino acid sequences, a consequence that would be lethal sooner rather than later! So such mutations would be strongly selected against during the course of evolution.

Messenger RNA links DNA and protein

DNA does not act alone in producing protein. While Gamow's ideas about coding were sound, his speculations that amino acids could somehow plug into cavities in the DNA double helix and assemble themselves into a protein went against all the experimental evidence.

When a cell grows, protein synthesis increases, and so does the amount of RNA (the 'other' nucleic acid) in the cell. DNA never moves out of the molecules, but protein synthesis occurs in the cytoplasm. It takes only a couple of minutes to make a protein molecule – and there is absolutely no evidence that DNA can leave the nucleus, visit the site of protein synthesis, and slip back undetected into the nucleus in this short time. There had to be an intermediary between DNA and protein.

RNA is this go-between. That RNA played a key role in protein synthesis had been suspected for many years. In 1954 Gamow had formed the RNA Tie Club, an informal association of (male!) scientists who wanted to understand how RNA helps to build proteins. Gamow ordered a set of ties with an RNA motif embroidered in silk on a black background. There were to be 24 members of the club – 20 for the amino acids, and four for the bases. Each member would have the symbols of the amino acid or base he represented on his tie. Gamow himself was the amino acid alanine (symbol Ala) and Crick was tyrosine (Tyr). The club's members dedicated themselves to bouncing ideas about coding and protein synthesis off one another. Throughout the 1950s formal reports, gossipy notes and letters of pure speculation about the puzzle of RNA flew around the world between Ala, Tyr, Val (Brenner was valine) and the rest of the club.

One of the main reasons why it took until 1960 to pin down RNA as the intermediary between DNA and protein is that three types of RNA exist in the cell – all with a different function. The RNA that acts as the intermediary is called messenger RNA (mRNA), but there is also transfer RNA (tRNA) and ribosomal RNA (rRNA).

It is mRNA whose concentration increases in the cell during protein synthesis. This is because the cell makes mRNA only when genes are being expressed – that is, making protein. mRNA is a copy of the gene that is being expressed. It passes out of the nucleus and into the cytoplasm, carrying the recipe for the protein with it, so that the amino acids can be assembled in the right order.

The copying of the gene into a strand of mRNA is called transcription. In many ways it resembles the replication of DNA – in particular, its driving force is also base pairing. First a small section of the DNA double helix of the gene to be expressed is unzipped by an enzyme called RNA polymerase (note that enzymes always have names that end with -ase). Next, RNA polymerase nudges single nucleotides – of which the cell has ample supplies – into position. But they only line up opposite one strand – not both – because this is transcription, not replication, and only one copy is going to be made.

At this point it is important to note that the two ends of a single DNA strand are slightly different chemically. One end is called the 5′ (5-prime) end, and the other the 3′ end. When two strands of DNA wind around one another in the double helix they do so 'head to tail' with the 5′ end of one facing the 3′ end of the other and vice versa. The single strand that the nucleotides now line up against is called the template, or antisense, strand. The other strand is the coding or sense strand.

The base-pairing rules are the same as in replication – with one exception. In RNA uracil (U) stands in for thymine, so wherever A appears in the gene sequence U, rather than T, will appear in the corresponding mRNA. Apart from this, the mRNA has the same base sequence as the coding strand, and both are complementary to that of the template strand. RNA polymerase – which is a very versatile enzyme – then clips the nucleotides together to make a strand of mRNA (Fig. 2.1).

mRNA is made a bit at a time. Short sections of the DNA double helix are unwound, transcribed, and wound up again – always moving from the 3′ to the 5′ end of the template strand. When RNA polymerase reaches a stop codon, the completed mRNA molecule slides away from the double helix, and the enzyme goes back to the beginning of the template strand ready to make the next mRNA molecule. RNA polymerase works at about the same rate as a photocopier – running off a single mRNA transcript in the time it takes to make a copy of a letter.

Transcription is such a vital process for all cells that any interference with it can have fatal consequences. More than 100 deaths occur every year from people eating a poisonous mushroom called the death cap (*Amanita phalloides*), which is found in oak and beech woods during late summer and autumn. It is easily mistaken for an edible species, but contains the toxin α-amanitin, which forms a tight bond with RNA polymerase, thus preventing it from carrying out its task. Most of the enzymes that carry out vital biochemical reactions in the organs of the body have a lifetime of just a few hours. Fresh supplies are always needed, and the supply

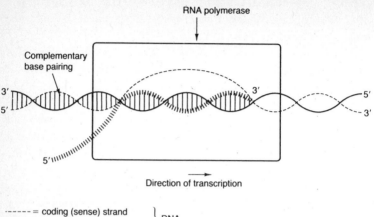

Fig. 2.1. Transcription of DNA. A strand of messenger RNA (mRNA) is formed with a base sequence complementary to that of the template strand of DNA. This process 'copies' information, in the form of the genetic code, from DNA to RNA.

chain starts with transcription of the appropriate genes. When α-amanitin invades the cells the effect is rather like imposing economic sanctions on a country that is wholly dependent on imports. When fresh enzymes are needed, none are forthcoming. The body's biochemistry is radically disrupted. Twelve hours or so after ingestion of the toxin, severe vomiting, diarrhoea, and stomach pains set in – followed by a symptom-free lull before catastrophe. Deprived of their vital enzymes, the liver and kidneys fail, and the circulation collapses. Even with intensive care, the toxin is fatal in around 10 per cent of cases. If there is no treatment, the death rate rises to 50 per cent.

Interference with transcription, however, can be turned to good effect. Rifampicin, an antibiotic used in the treatment of tuberculosis, acts by interfering with the fastening together of mRNA chains by RNA polymerase molecules of the infecting bacteria. Fortunately, human RNA polymerase molecules differ sufficiently from the bacterial version to remain unaffected by the antibiotic. The anti-cancer drug actinomycin D sits on double-stranded

DNA so that it cannot be unwound ready for transcription. The drug can be targeted against rapidly dividing cancer cells because these need to make a great deal of mRNA.

From mRNA to protein

In 1955, Crick and Brenner circulated the RNA Tie Club with a lengthy note in which they argued that there must exist, in the cytoplasm, small molecules that act as 'adaptors' – physically linking the codons in mRNA with the amino acids for which they code. After much experimental effort, it was discovered that the much sought after molecules were another sort of RNA – transfer RNA (tRNA).

There are tRNA molecules for each of the 20 amino acid used to build proteins. The first tRNA base sequence – from yeast – was worked out by Robert Holley and his team at Cornell University in 1965 and the rest soon followed. Each tRNA is a single strand of RNA containing between 73 and 93 nucleotides. Internal pairing between some of the bases pulls the molecule into a clover-leaf shape. The stem of the clover leaf ends in a three nucleotide hook, which can pick up an amino acid. So this part of the molecule 'plugs into' the amino acid. The loop opposite this stem has a base triplet called an anticodon. This section 'plugs into' mRNA because the anticodon sequence is complementary to the codon of the amino acid that can be picked up by the hook. tRNA therefore is a beautiful piece of molecular design, which brings together mRNA and the components of the protein it codes for.

For instance the tRNA that carries the amino acid phenylalanine has an anticodon with the sequence AAA. This can base pair with UUU on mRNA (remembering that U stands for T in RNA). Working backwards, UUU was copied from AAA on the template strand of the gene during transcription. So when codon and anticodon meet, tRNA brings a phenylalanine molecule ready to be assembled into a protein molecule – exactly what the gene asked for!

Proteins are actually assembled, one amino acid at a time, by a molecular machine in the cytoplasm, of which mRNA and tRNA are vital working parts. This machine is housed in the ribosome – a particle made of protein and a third kind of RNA, ribosomal RNA (rRNA). There are thousands of ribosomes scattered throughout the cytoplasm, all busily turning out copies of protein molecules.

The ribosome consists of two subunits that fit together like the top and bottom of a squashed figure-of-eight. The subunits snap together over a molecule of mRNA, trapping it in the gap between them. The protein translation machine is now ready to start (Fig. 2.2). Put simply, the mRNA molecule moves through the ribosome, exposing its codons one at a time to the anticodons on the amino acid loaded tRNAs, which are hovering in the vicinity. As soon as they detect 'their' codon, they move into position.

This process lines the amino acids up – two at a time – in the right order. An enzyme that specialises in this sort of work unhooks them from their respective tRNAs and hooks them together. The liberated tRNAs then drift off to load up with fresh supplies of their amino acids to keep the protein translation machine going. Then the mRNA moves along the ribosome, and the next tRNA moves into position, ready to be fastened into the growing protein chain.

Each mRNA molecule might be translated over and over – producing thousands of protein molecules. Electron microscopy has captured the beauty and economy of the process, showing how several ribosomes work on one mRNA strand at any one time at different positions along its length.

Needless to say, interference with the finely tuned machinery of protein translation can wreak havoc in the cell. Before the introduction of effective immunisation, diphtheria was one of the most feared diseases because of its high fatality rate. Diphtheria is caused by the action of a toxin produced by some strains of the bacterium *Corynebacterium diphtheriae*, which inhabits the upper respiratory tract. The toxin disables the enzyme, translocase, that normally moves the mRNA along the ribosome, forcing protein synthesis to stall at a very early stage. What makes this toxin especially deadly is that it does not itself block the enzyme, but acts as a catalyst, forcing a cell component called ADP-ribose to do the job instead. So a single toxin molecule can kill a cell, because it can visit every ribosome, forcing ADP-ribose to put the translocase

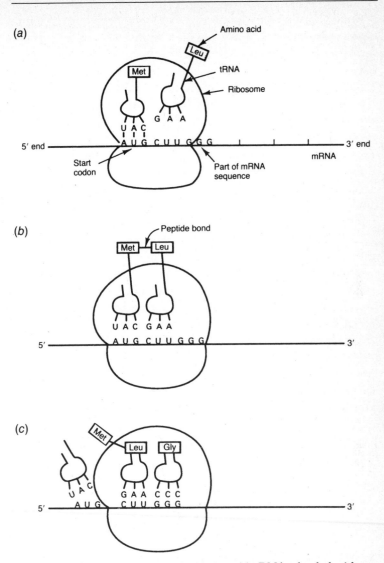

Fig. 2.2. Stages in protein translation. (*a*) tRNAs, loaded with amino acids, move into position on the ribosome as codon–anticodon pairs are formed. (*b*) A chemical bond, called a peptide bond, is formed between two amino acids. (*c*) Protein assembly continues as unloaded tRNAs drift away, and the next amino acid moves into position.

enzyme out of action at each site. No wonder one microgram of diphtheria toxin can be lethal to an unimmunised person.

Many antibiotics dispose of bacteria by interfering with bacterial – but not human – ribosomes. For instance, erythromycin, which is useful in treating bacterial infections in people who are allergic to penicillin, effectively jams the mRNA in the ribosome so it cannot be 'read'. Streptomycin, one of the first and most effective weapons against tuberculosis, stops the first tRNA from approaching the ribosome – so that bacterial protein synthesis closes down.

A new approach to blocking protein synthesis is to disable mRNA before it even arrives at the ribosome. This can be done by using a short stretch of single-stranded DNA complementary to part of the mRNA sequence. Such a molecule is known as an antisense oligonucleotide (oligo for short) and it is readily synthesised in the laboratory. The 'antisense oligo' captures the mRNA and forms an RNA–DNA double helix, which cannot be used by the ribosome. So no protein synthesis can occur.

Antisense technology has given us tomatoes with a longer shelf life, and fashionably pale petunias (see Chapter 10) as well as new insights and potential therapies for cancer, AIDS and other viral infections. Already antisense oligos are being tried out in leukaemia patients, in the hope that they will be able to 'turn off' the harmful activity of some of the genes involved in the development of cancer (see also Chapter 8). Test tube studies suggest that antisense oligos can block the translation of the genes that help HIV and influenza viruses to reproduce.

Proteins are special

We have now seen how DNA makes proteins. Before exploring DNA further, it is time to put proteins back in the spotlight to see why the DNA–protein connection is so important. As we saw in Chapter 1, biochemists used to think proteins were more important than DNA. Nowadays, the balance has swung in the other direction. In reality such divisions are artificial. Both nucleic acids and proteins are vitally important in sustaining life; other biologically active molecules play a supporting role.

The crucial role played by proteins is not readily apparent from everyday information about health and nutrition. For instance – we need to eat a balanced diet of carbohydrates, proteins, fats, minerals, vitamins and fibre – all washed down with copious amounts of water (and of course, we can get by for far longer without food than we can without water). This advice tells you nothing about what these molecules do in the body. Once in the digestive system the carbohydrates, proteins and lipids (fats) are broken down into smaller molecules – glucose, amino acids and fatty acids and these act as the 'raw materials' of the cell.

Then the cell builds the molecules it needs from these raw materials – hormones, enzymes, nucleic acids, and so on. But enzymes – which are themselves proteins – are absolutely essential to this building work. Think of the conditions inside the cell – a temperature of 37°C, a watery environment – and then ask any organic chemist if he or she could build the molecules needed to sustain life from the cell's raw materials under these conditions, but without the enzymes. The answer would be a resounding 'No'. Besides enzymes there are proteins that are used as building materials within the body: actin and myosin in muscle, collagen in bone and keratin in skin.

Once a protein has been synthesised, it goes about its business but sooner or later it will come to the end of its working life. The corresponding gene will have to turn out some new copies to replace it. The picture that emerges is one of genes and proteins operating together; the genes supply the proteins, and the proteins attend to the biochemical 'work' needed to keep the cell, and ultimately the whole organism, going.

New information channels – retroviruses and prions

The Central Dogma – information flows from DNA to RNA to protein – has been at the heart of molecular biology for over 40 years. Once the molecular details of gene expression, as described above, had been clarified, the impossibility of such a complex process running backwards seemed obvious. Information, Crick said, could never flow from protein back to DNA.

However, the Central Dogma does allow certain other information channels. Obviously information passes from one DNA strand to another during replication. The flow of information from DNA to RNA can also be reversed. This happens in a group of viruses known as the retroviruses. The human immunodeficiency virus (HIV), which is associated with AIDS, is a retrovirus, as are some of the viruses that cause tumours in humans and other animals.

Viruses are the simplest of all life forms. They consist of a single or double strand of nucleic acid (DNA or RNA) surrounded by a protein coat. They have no independent existence but must invade a host cell and take over its cellular machinery in order to express their genes. The genetic material of a retrovirus is single-stranded RNA. It replicates inside its host cell by copying its RNA-based genes into a strand of DNA (the reverse of normal transcription, which makes an RNA copy of DNA). An RNA–DNA double strand is formed by this process. Then the RNA strand breaks down and the DNA strand doubles up – all under the influence of enzyme action, of course. Now the viral genes are in the same form – DNA – as the genes of the host cell, which innocently adopts them and treats them just like its own genes.

The viral enzyme that catalyses all three steps of the copying of RNA into DNA is called reverse transcriptase (RT) and it was discovered independently by Howard Temin and by David Baltimore in 1970. The discovery provoked an astonished editorial in the journal *Nature* with the headline 'Central Dogma reversed'.

The drugs currently used against AIDS – AZT, ddI and ddC – act by blocking the action of the RT enzyme of HIV and so are known as RT inhibitors. The strategy is fine – but these particular drugs have proved to have severe limitations in the treatment of AIDS, so there is an urgent need for better RT inhibitors. By the end of 1991, Thomas Steitz and his team at Yale University had solved the three-dimensional structure of HIV RT using X-ray crystallography. This reveals, in great detail, the part of the enzyme that needs to be blocked off by an inhibitor. Now the pharmaceutical industry can push forward with the search for a tailor-made molecule to do the job – searching through computer libraries of known molecular structures for likely candidates.

Over the last ten years, the Central Dogma has been challenged yet again – by the heretical notion that information could, under certain circumstances, pass from protein to protein without any nucleic acid involvement. Infectious diseases are generally caused by the invasion of a host by a microbe – bacterium, virus or fungus – which then replicates using nucleic acid, damaging its host in various ways as it does so.

However, there is a group of diseases that affect the nervous system of humans and other animals in which the infectious agent does not appear to be a microbe. Instead it appears that the disease is transmitted by a protein-containing 'particle' called a prion. Scrapie, a fatal, nervous disease of sheep and goats, falls into this category, as does bovine spongiform encephalopathy (BSE or 'mad cow disease'), which has killed more than 70 000 cattle in the UK since 1985. BSE is caused when cattle consume tissue from sheep infected with scrapie as part of their feed.

Prion diseases that affect humans include kuru and Creutzfeldt–Jakob disease (CJD). Kuru has been linked to the handling and consumption of human brain tissue by the Fore people of Papua New Guinea during funeral rites (thankfully the disease, which used to affect up to one per cent of the population, is disappearing as cannibalism is abandoned).

CJD is a rare fatal dementia; cases have arisen among people operated on by prion-infected instruments, or after treatment with contaminated human growth hormone. This hormone, which is used to treat dwarfism, is now made by genetic engineering (see Chapter 5) but prior to this was extracted from the pituitary glands of human corpses. Mortuary technicians were paid a nominal sum to collect the glands. Very occasionally a gland was taken inadvertently from the corpse of a CJD victim – and so the infection was passed on.

The infectious nature of the protein (known as prion-related protein or PrP) in prion diseases has been demonstrated by experiment. When an extract of infectious tissue was treated with chemicals that destroy protein it showed a decrease in infectivity.

But when the tissue was exposed to ultraviolet light, which inactivates nucleic acid, it remained as infectious as before. This suggests that prions do not contain nucleic acids. Note how similar this strategy is to Avery's experiments, which were discussed in Chapter 1.

The prion hypothesis – the idea that PrP can reproduce itself without nucleic acid – is still controversial. Some scientists argue that the prion must contain nucleic acid that is, for some reason, evading detection. Maybe the molecules are very small, and are in some way unusual in their structure and are not destroyed by ultraviolet light in the above experiments.

PrP turns out to be an abnormal variant of a protein produced naturally by the body. Stanley Prusiner of the University of California, a pioneer of prion research, has suggested the following mechanism for prion replication. An abnormal PrP molecule attaches itself to a normal one. The association converts the normal molecule into an abnormal one – a transformation that is accompanied by a significant alteration in the shape of the molecule. The two molecules then separate. Each of the two now abnormal molecules could then repeat the process – resulting in four abnormal molecules. This corruption would, like the multiplication of bacteria, become a cascade of repeated doublings. Large numbers of abnormal PrP molecules appear to form characteristic lesions known as plaques in brain tissue, which lead to the symptoms of prion diseases. Scientists at the Massachusetts Institute for Technology and the US National Institutes of Health have shown that abnormal PrP can indeed turn normal PrP 'bad' – at least in the test tube, suggesting that Prusiner may well be right.

The investigation of prion diseases is important. Greater understanding in this area will help to clarify whether BSE can be transmitted to humans. Knowledge of how prions destroy brain tissue might also have implications for other neurodegenerative conditions such as Alzheimer's disease, where plaques similar to the prion plaques are found, and Parkinson's disease.

Switching gene expression on and off

There are certain proteins that must be available at all times for cells to keep functioning. For instance, histone proteins are vital so that cells with nuclei can organise their DNA into chromosomes, as we shall see in Chapter 3. Other proteins are more specialised, and may be needed only at a certain stage of development, or by certain tissues. An example is the hormone insulin, which is manufactured in the pancreas, and used to control the amount of glucose in the blood (blood sugar). Insulin is not produced, or used in, the brain, lung or any other tissues.

However, every cell in the organism has the potential to express genes for both essential and specialised proteins. This must be so, because the DNA of the original cell of the organism (the fertilised egg in humans, for example) is copied into each cell as the organism grows and therefore contains a full complement of genes.

The genes for essential proteins are known as housekeeping or constitutive genes and these are expressed all the time. Inducible genes, which code for specialised proteins are turned on only when and where expression is required, otherwise the cell would waste precious biochemical energy turning out proteins that were not actually required. Worse, the unwanted proteins might interfere in the cell's biochemistry. Such a scenario would not be conducive to the organism's survival. If the insulin gene were left switched on all the time, for instance, the hormone would have a devastating effect on blood chemistry that would be fatal within a short time. All the blood sugar would end up being stored in the liver instead of fuelling vital organs such as the brain.

Gene expression is regulated by 'on–off' switches. There is more to a gene than the piece that is copied in transcription. Remember that this runs in the direction 3' to 5'. It turns out that there is DNA 'upstream' from the 3' end – just before the start of the sequence that RNA copies – which acts as the switch. The first example of one of these switches was discovered by François Jacob and Jacques Monod, working at the Pasteur Institute in Paris in the late 1950s and early 1960s. They started by exploring the way in which bacteria change their pattern of gene expression in response to changes in their environment. All organisms require fuel to power cellular activities. For most organisms this fuel is a source of carbon, such as the sugar glucose. Bacteria can use a far wider variety of carbon sources than other organisms can. Some even feed off chemicals that we would regard as highly toxic, such as hydrocarbons and coal tar (and, as we shall see in Chapter 11, this ability can be put to good use).

The bacterium *E. coli* can use lactose – a sugar that occurs in milk – as a food source when it is available. The enzymes required to break down lactose to extract the energy from it are different from those required to break down other sugars such as glucose. One of these enzymes is known as β-galactosidase (β-gal for short). When lactose is absent, fewer than ten molecules of β-gal can be found in a single bacterial cell. But if you place the bacteria in an environment that does contain lactose, by adding some milk powder for example, the number of β-gal molecules jumps to around 3000 per cell.

Jacob and Monod directed their efforts towards working out how the bacteria could respond so 'intelligently' to a chemical signal (the presence or absence of lactose) in their environment. They studied mutant strains of *E. coli* that were defective in their ability to synthesise β-gal. Of particular interest to them was one mutant that poured out large amounts of the enzyme whether or not lactose was present in the bacterial culture medium. Somehow the mutant had thrown off the constraints on expression of this gene and turned it into a constitutive gene. The mutant was not producing some substance that normally regulated the gene expression, Jacob and Monod argued. So the gene switches were on all the time – rather like a central heating system that burns up fuel all the time when its thermostat is broken.

The next stage in one of the most elegant series of experiments in the history of molecular biology was to try to repair the genetic defect in the constitutive mutant. Jacob and Monod carried out what is known as a complementation test. They 'mated' the constitutive *E. coli* mutant with another mutant of the same species. This one had normal regulation, but had a defect in the β-gal gene itself that prevented the enzyme being produced. The 'offspring' of this union produced the enzyme, but only in the presence of

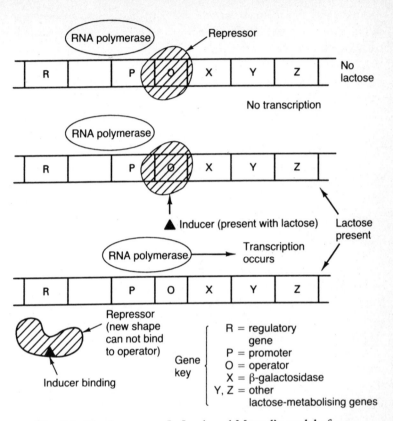

Fig. 2.3. The *lac* operon. In Jacob and Monod's model of gene control, RNA polymerase is unable to transcribe the gene because the repressor protein prevents access. The inducer acts as a chemical signal turning on gene expression by removing the repressor and allowing transcription to go ahead.

lactose. Jacob and Monod concluded that the β-gal gene in the constitutive mutant had been brought under control – by the regulatory substance being produced by the second mutant. The enzyme could not have been produced by the second mutant because its gene for this enzyme was defective.

These results led to Jacob and Monod's formulation of the operon model of gene expression (Fig. 2.3), and a well-deserved

Nobel Prize for the two men in 1965. We can now expand our picture of the gene into a stretch of DNA made up of a coding region, which specifies the amino acid sequence of a protein, and a control system.

Upstream from the first codon of the coding region is the gene's 'signal box'. This is a short stretch of DNA consisting of two sections: the promoter and the operator. RNA polymerase sits on the promoter region, ready to transcribe the gene that follows. The operator lies between the promoter and the first codon. Jacob and Monod suggested that in the so-called *lac* operon, which contained the genes for breaking down lactose, a regulatory substance was bound to the operator. This blocks the access of RNA polymerase to the coding region, so no mRNA can be made. So the *lac* operon's default status is 'off'.

When lactose is present in the surroundings of the bacteria, it is rapidly converted to a closely related substance called allolactose. This attaches itself to the regulatory substance and pulls it off the operator. There is now no barrier to transcription; RNA polymerase moves into position and begins to transcribe the coding DNA sequence. In this state the operon is 'on'.

Later experiments led to the discovery of the regulatory substance, which turned out to be a protein. The shape of this so-called repressor protein is ideal for binding to the operator DNA; but when the allolactose molecule binds to it, the shape of the repressor protein alters to accommodate it. This structural shift affects the whole protein molecule, and results in it losing its grip on the operator, i.e. giving the green light to RNA polymerase. Together, the gene for the repressor protein, the promoter and the operator make a sophisticated control system for gene expression.

Monod was not impressed. A life-long communist who marched with protesting students in Paris in 1968, he saw the *lac* operon as proof of life's intrinsic lack of meaning. Other scientists might have seen the hand of a Creator in the beauty and seeming intelligence of this bacterial control system. For Monod it proved the power of chemical bonding over the futility of the anthropic principle, which, put simply, suggests that there is some hidden purpose behind the evolution of life. His reductionist views are discussed further in Chapter 12.

We now know that regulation of transcription is a major method of orchestrating gene activity in all organisms, and it makes sense to nip gene expression in the bud in this way, so that no biochemical energy is used in making redundant RNA transcripts. The β-gal gene is one that has to be turned on by an inducer – in this case allolactose – because its default status is 'off'. Some genes are controlled in the opposite way; their default status is 'on', and they have to be turned off. Again it is a protein that turns the switch.

The amino acid tryptophan (Trp) – the ingredient in milky bedtime drinks that is supposed to make you feel sleepy – is manufactured by bacteria, with the aid of several enzymes. However, there is little point in a bacterium continuing to turn out tryptophan once it has made enough for its biochemical needs. Tryptophan controls its own synthesis in a scenario that is the mirror image of that of the *lac* operon. The *trp* operon, which contains the genes for the enzymes that make the amino acid from raw materials in the cell, is under the control of a 'repressor' protein. This binds to the operator only when there is plenty of tryptophan around. Tryptophan sticks to the repressor protein, moulding it into a shape that fits the operator and blocks transcription.

Gene expression in other organisms – from yeast and fruit flies to humans – is also largely controlled at the level of transcription. The control machinery is more complex though, and is currently the subject of intense study in laboratories all around the world. Many of the key players have been identified: sequence motifs in the promoter region; regulatory proteins known as transcription factors, which help RNA polymerase to transcribe genes; and so-called enhancer sequences, sections of DNA that could be thousands of base pairs away from the promoter yet exert a 'remote control' effect on the gene.

Transcription factors act as molecular keys that turn on transcription by fitting into grooves in promoter DNA. For instance, the combination of two helices at right angles to one another has been found in many proteins regulating the genes involved in mammalian development. The helix–turn–helix motif in these proteins acts like a finger and thumb that can grab hold of the promoter and turn on the gene.

Another important pattern found in transcription factors is the zinc finger. This is a protruding segment of amino acids enclosing zinc atoms. Zinc is a metal used as a coating on iron to protect it from rust, as well as being blended with copper to form the alloy brass. It is also an essential trace element in the diet (good sources are oysters, ginger, nuts and seeds) because it is needed in some enzymes, and in zinc fingers. The first zinc fingers were discovered in a transcription factor from a frog in 1985 by Aaron Klug and his team at the Laboratory for Molecular Biology in Cambridge. Now over 200 transcription factors containing zinc fingers are known. Sets of zinc fingers insert themselves into the grooves between the turns of the DNA double helix in the promoter region. They are so important that one per cent of human DNA codes for proteins with zinc fingers.

Some transcription factors do not act alone. Instead they function as a bridge between DNA and hormones. These powerful molecules exert a diverse range of physiological effects, from the development of sex organs in puberty in humans to the ripening of fruit and the falling of leaves in the autumn. Hormones such as oestrogen – the female sex hormone – bind to transcription factors, and only then will the zinc fingers curl around their DNA target. A lack of zinc in the diet can actually impair sexual development, because the transcription factors cannot function without it, and the sex hormone cannot turn on the appropriate genes without transcription factors.

Sometimes this hormone-operated switch goes wrong. If the hormone turns on genes which produce proteins that cause cells to divide inappropriately, a cancer may result – or an existing tumour become more active. Many breast and prostate cancers are hormone sensitive in this way, responding to oestrogen or testosterone, respectively. This discovery has led to a new drug for breast cancer, tamoxifen. This is a so-called anti-oestrogen: it blocks the site on the transcription factor that is normally occupied by oestrogen. Tamoxifen has been successful in the treatment of a significant number of breast cancers, and is being tried as a preventative drug for women who are at high risk for developing the disease (see also Chapter 7).

A surprising number of chemicals in the environment can mimic the action of oestrogen. For instance, crocodiles and fish exposed to these oestrogen-like substances develop 'feminised' hormone profiles, with male fish producing female egg-yolk proteins. These chemicals come from a variety of sources, such as the breakdown of detergents and pesticides. More controversially, they have been implicated in an apparent fall in human male sperm count.

The good news is that there are natural anti-oestrogens in the plant world. One clue to the puzzle of why women from South-East Asia, Japan and China have a low rate of breast cancer seems to be their high consumption of soya bean products such as tofu and soy sauce. Soya contains isoflavones, which, like tamoxifen, dilutes the effects of oestrogen.

Gene expression can also be controlled to protect us from the effects of environmental pollution. Heavy metals such as cadmium can trigger the transcription of the metallothionein gene. This codes for a small protein that binds tightly to heavy metals and prevents them from harming the cell. Thus, the metal triggers the production of what is essentially a detoxifying system.

Control of gene expression underlines the intimate links between genes and proteins. Proteins control genes, which in turn control the production of proteins (the whole range, not just those that bind to DNA). In proteins it is the shape of the molecule that provides its function. In DNA, sequence is more important. The base sequences of genes determine amino acid sequences of proteins, which are a major factor in determining their shape. The more we find out about genes and proteins, the more the subtle ways they work together are emphasised – each being driven by unique features of their chemical structure.

3

A close up of the genome

The sum total of the DNA in an organism is known as its genome. Although DNA was discovered within the nucleus, this is not the only place where it is found in cells. Some cells do not even have a nucleus.

As far as molecular biology is concerned the distinction between having a nucleus and not having one turns out to be rather important, and is used for classifying organisms at a cellular level. A daisy and a giraffe have hardly anything in common as far as outward appearance is concerned, but to a molecular biologist they are similar because they are both made of cells that have nuclei.

Organisms like this are called eukaryotes. Their DNA is found in the nucleus, but also in cellular structures called mitochondria and chloroplasts. These are the sites of important biochemical activity. Energy production takes place in mitochondria, while the synthesis of glucose from carbon dioxide and water in the presence of sunlight (photosynthesis) happens in the green chloroplasts of plant cells. The fact that mitochondria and chloroplasts have their own DNA suggests that they may have been free-living organisms earlier in evolution (the significance of this is discussed further in Chapter 4).

Organisms whose cells do not have a nucleus (or at least not one surrounded by a membrane like the eukaryotic nucleus) are called prokaryotes. Their DNA lies free in the cell, usually in a closed loop. In addition to this, some bacteria contain small circles of DNA called plasmids.

The C-value paradox

The C-value is a measure of the amount of DNA in a given organism. It is measured in picograms of DNA in a single cell (a picogram is a million millionth of a gram). So to get the total mass of DNA in the organism you would, of course, have to multiply the C-value by the total number of cells present. Our model prokaryote, *E. coli*, has a C-value of 0.004 picograms, for example, while dogs, horses and mice have C-values of around 4 picograms.

The bigger the C-value, the longer is the DNA sequence in the organism's genome. In *E. coli* that 0.004 picograms gives us a genome that is a DNA molecule with four million base pairs (or bp). Because DNA is a double helix, molecular biologists like to talk in base pairs or better still kilobase pairs (kbp) (as even the simplest genomes contain well over 1000 bp). Four million base pairs gives you a DNA molecule which is 1.36 millimetres in length.

Now take a model eukaryote – yeast, for example. Its genome has 13 500 kbp. If you were to remove all the DNA from its nucleus (where it is wound up in the chromosomes – of which more later) and piece it together, it would be 4.6 millimetres long.

Yeast is a more sophisticated creature, biochemically, than *E. coli* so it is hardly surprising that it has a bigger genome. Extending the argument, you might expect the human genome to be the longest of all. Humans have around three million kbp of DNA in each cell – about one metre in total length and nearly 1000 times more than *E. coli* (here we are ignoring, for the moment, the fact that humans have paired chromosomes and we are merely looking at the DNA in a set of single chromosomes, which is the convention in genetics when talking about C-values). But the waterlily, salamander, South American lungfish, and the common house plant *Tradescantia* have between 20 and 70 times as much DNA as humans.

There are also some startling variations in DNA content between related species. A toad may have as much DNA per cell as humans, but a frog may have twice as much. Genome size in amphibians varies between 700 000 kbp and 100 million kbp.

Evidently there is no simple relationship between the apparent complexity of an organism and the amount of DNA in its genome. This discrepancy is known as the C-value paradox. So far we have seen that genes are made of DNA, but does all DNA contain genetic information? A sample calculation shows that it does not. The DNA 'molecule' in a human cell contains three million kbp (imagine the DNA from a single set of chromosomes pieced together into one long molecule). If we say that the average protein has around 500 amino acid residues (i.e. building blocks), then the coding sequence of its gene has 1500 base pairs (remember that each codon has three bases, meaning that 500 amino acids will need 500×3 bases to code for it). Add a generous 500 base pairs for regulatory sequences, to make around 2 kbp per gene. Divide this into the three million kbp of the whole genome and this gives us between one and two million genes – if the whole genome does in fact code for protein (i.e. if it is nothing but genes).

Do we really need so many genes to account for our biological complexity? From what we know of the enzyme requirements of our cells the answer has to be 'no'. Current estimates suggest that 50000 to 100000 genes is probably quite sufficient to provide a human with all the protein necessary to sustain his or her bio-chemical life. This represents around two per cent of the genome.

Some of the rest can be accounted for. For example, it turns out that many eukaryotic genes contain many more than 2 kbp, because of their complicated structures. Some genes are present as multiple copies, and there are various other bits and pieces of DNA whose possible origin and function is known – or at least suspected by biologists working in this area. These features are discussed in more detail later in this chapter.

This still leaves vast tracts of eukaryotic DNA of no known function (while prokaryotic genomes are remarkably compact). This DNA is often called 'junk' DNA. Some scientists, such as the British biologist Richard Dawkins, have suggested that the existence of 'junk' DNA – and the variations in C-value within and between species – might be explained by the concept of the selfish gene.

Dawkins' ideas are discussed in more detail in Chapter 12 but, briefly, he proposes that the replication of DNA is the driving force of evolution. DNA replicates, whether it is of use to the organism or not, and it expands to the limits tolerated by a particular organism. So selfish (or junk) DNA replicates alongside DNA that codes for protein. The protein-coding DNA – the genes – create an organism that the selfish DNA merely uses as a vehicle for its own survival.

Packing and parcelling – fitting DNA into cells

Organisms with a great deal of 'junk' DNA are like holidaymakers trying to fit 'everything but the kitchen sink' into their suitcases when it comes to trying to stuff their genomes into their cells. Even bacteria, with their streamlined genomes, have had to evolve efficient packing strategies to pack their DNA into their cellular suitcases.

A bacterial cell has a diameter of around one micrometre (a micrometre, μm, is a millionth of a metre or one thousandth of a millimetre), and its DNA molecule, which is equivalent to the entire genome, is around a thousand times longer than this. If it was left as a single loop it would never fit into the cell. In 1963 Jerome Vinograd discovered that looped DNA can exist in a 'supercoiled' form inside the cell, where the sides of the loop are further twisted around one another. You can see the compacting effect of supercoiling with a simple experiment using a loop of string, a nail, and a pencil. Hook the loop round the nail (you can use a door handle in the absence of nails) and pull it taut, horizontally. Now insert a pencil into the other end of the loop. Keeping the loop taut, rotate the pencil so the two sides of the loop wind round another over and over again. If you now loosen the loop, it will spring into a supercoiled state. Compare this to the original length of the loop!

In eukaryotes, DNA is packed away in chromosomes (Fig. 3.1) – the thread-like structures within the nuclei that were discussed in Chapter 1. Chromosomes are made up of a material called chromatin, which is composed of DNA and proteins known as histones. Think of the histone proteins as roughly spherical molecules,

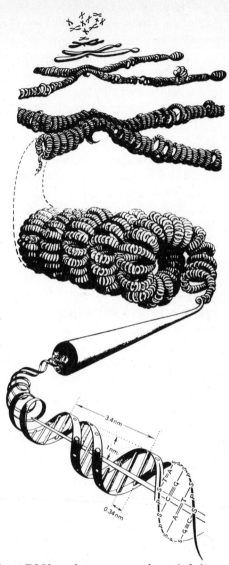

Fig. 3.1. From DNA to chromosomes. An artist's impression of the packaging of DNA in chromosomes. The chromosomes are shown at the top of the picture, and gradually 'unravelled' to reveal the DNA molecule. The scaffolding proteins are not shown.

around which you could wind DNA, like cotton on a reel. The structure of chromatin is quite a feat of packaging. It consists of a number of bead-like structures called nucleosomes, linked together like a necklace. If we take a nucleosome apart, we find two turns of supercoiled DNA wound onto a core of eight histone molecules. This takes care of 200 base pairs of DNA, reducing them to one seventh of their original length. Then the nucleosome necklace is coiled up into a thick fibre – a process that reduces the length of the DNA molecule by an overall factor of 40.

This must be followed by further coiling and looping, because the chromosomes are around 0.3 millimetres in total length (if you took all 46 and laid them end to end), while the DNA they house has a total length of just over two metres (note that this time we are looking at *all* the chromosomes, whereas before we were looking at the DNA in a single set, which had a length of just over a metre) – thus packing compacts the DNA by a factor of approximately 10 000! Electron microscopy has revealed some of the details of this elaborate chromosomal architecture. If the histone proteins are stripped away from the chromosomes, the nucleosomes fall a part to leave long loops of DNA surrounding a scaffold made of proteins that are not histones. This protein scaffold provides the support needed for the sculpting of the chromatin fibres into their final compact form in the chromosomes.

The number and size of chromosomes varies between species. In most cells they occur in pairs; such cells are said to be diploid. However, germ cells (like the sperm and egg) contain just a single set of chromosomes and are called haploid. When two germ cells come together in the fertilisation of plants and animals a diploid cell is created as the single chromosomes pair up. Generally C-values are discussed – as mentioned previously with reference to the amount of DNA in a haploid cell, because this contains all the organism's genes. The significance of sexual reproduction, where genes from two different haploid cells are mixed, as described above, is discussed in Chapter 7.

The best time to look at chromosomes under a microscope is just before a cell divides. Each chromosome is made of two thread-like structures called chromatids, joined in a central region known as the centromere. When the cell divides, the chromatids – which are exact copies of one another – peel apart and one goes into each new cell. In 1970 scientists at the famous Karolinska Institute in Sweden found a new way of staining human chromosomes that revealed a pattern of dark and light bands. This led to a system of naming different bits of the chromosome, which was to prove useful in constructing genome maps. The centromere divides the chromatids into two unequal sections. The shorter is known as the 'p' arm and the longer as the 'q' arm. Then each arm is divided into one, two, or three numbered regions, working out from the centromere. The bands in these regions are numbered too. This means we can give map references to significant places in the genome. For instance, the map reference for the human insulin gene is 11p15, meaning it is found on chromosome 11, on the shorter p arm, in the fifth subdivision of the region closest to the centromere.

Charting the landscape of the genome

As DNA technology has advanced it has become possible to scan genomes from a wide range of organisms for features of interest. Gradually, a geography of the genome is emerging. These efforts have been brought to a focus by the various genome mapping projects that have been started, since the late 1980s, in laboratories around the world.

The aim of these projects is to produce maps of various types and scales. The most detailed of these would be the actual base sequence of a genome from start to finish, which would be rather like producing a map of a country that gave the location of each house. Although this is the eventual aim of the genome mappers – and it is close to completion for *E. coli* – it is likely to be some years before this is achieved for any eukaryotic organism.

Genome mapping does not mean starting at the first base of a DNA molecule and patiently working your way through millions or even billions of bases till you reach the end. Instead, it involves a broad brush approach, where the genome is chopped up by enzymes into DNA segments of a manageable size. Then each piece is scanned by a set of DNA probes. These are short lengths of DNA tagged with either a fluorescent or radioactive atom or molecule that flags the probe's complementary sequence on the genome. Scanning is done by adding the probe to the DNA mixture. The probe seeks out its complementary sequence and sticks to it, marking that location with a 'flag'. So we could use a probe of sequence AAAAAAAA, tagged with radioactive phosphorus, to give a radioactive 'flag' everywhere in the genome where TTTTTTTT is found because the probe will seek out and bind to this complementary sequence by base pairing. Then the scanned pieces of DNA have to be put back together in some sort of order – a very demanding task! One powerful approach relies on looking at regions that overlap two neighbouring fragments. Here is a simplified example of how this works. Suppose we have three DNA fragments A, B and C with different 'flag' patterns and we would like to know if the order is ABC, BCA or CAB. The trick is to chop up the DNA again – but with a different set of enzyme tools that will give you the overlapping regions between the three fragments. Suppose the true order is CAB. This should be revealed by the CA overlap having a 'flag' pattern characteristic of C and A rather than A and B. Patient piecing together of fragments in this way will give a genome map where at least some of the features are marked out. Interesting regions are then explored more fully.

There was enormous – but not unanimous – enthusiasm for the genome projects when they were first launched, but the type of experimental work described above is tedious in the extreme for those who have to do it. This led Sydney Brenner, whose role in cracking the genetic code was described in Chapter 2, to remark that the task of mapping human DNA could be like a vast prison for scientists. He joked that offenders could be allocated a 'stretch' of DNA to map – its length depending on the severity of the crime. Minor offences such as borrowing a colleague's bottles of enzyme without asking would merit a kilobase of DNA, while scientific fraud would attract a punishment of a whole chromosome to analyse.

Increasingly, though, machines and robots are coming to the rescue and carrying out much of the repetitive work involved in genome mapping. The floods of information that are beginning to emerge are being channelled into massive computer databases and made accessible to genome mappers around the world (although some companies are now beginning to argue for some restrictions, to protect their investments and possible applications such as testing for disease genes).

All this work aims to produce two different kinds of genome map. The first shows the order of the genes on a chromosome. This form of map was pioneered by the American scientist Thomas Hunt Morgan in the early twentieth century. Morgan rose to the challenge of synthesising the ideas of Darwin and Mendel and began to look at patterns of inheritance in the fruit fly (*Drosophila*), which, as the fly's breeding cycle is two weeks, gave results more quickly than Mendel's pea plants.

Morgan noticed that there was a tendency for some genes to be inherited in groups. For example, having white eyes and being male usually went together. Soon he had sorted the fruit fly genes into four different groups. It turns out that the reason why these groups are often inherited together is that they are physically linked – by being on the same chromosome. So each group is called a linkage group. This is true of other species – the number of linkage groups equals the number of chromosome pairs (four in fruit flies, 23 in humans and so on).

As he pressed on, Morgan observed something else. Linkage was never complete but varied between 50 and 100 per cent, depending on how far apart two genes were on the chromosome. The further apart they were, the more they acted as if they were on separate chromosomes, and the closer the two genes were, the more they acted as one.

By 1915 Morgan had constructed a linkage map that ordered 85 fruit fly genes on the four chromosomes. Seven years later there were 2000 genes on this map. The fact of incomplete linkage means that chromosomes are not indivisible units. What happens is that chromosomes are shuffled when germ cells (sperm and egg) are made by a specialised process of cell division called meiosis (Fig. 3.2).

Meiosis starts when the paired chromosomes come into close contact and an enzyme nicks each of the four chromatids at corresponding points. Supposing we label the two chromosomes A

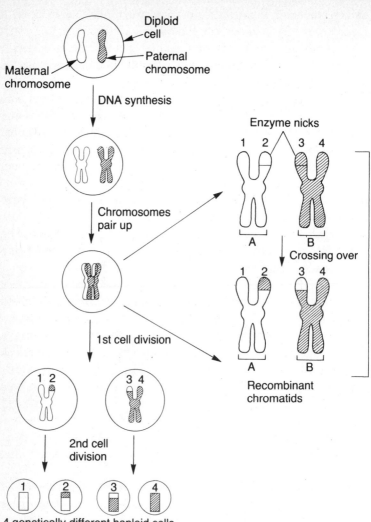

Fig. 3.2. Meiosis, crossing over and recombination. In meiosis there are two cell divisions and the end result is four new cells. During the first cell division, genetic material is shuffled between the chromosomes. The second cell division does not double up the chromosomes, so each of the four cells has unpaired chromosomes, all with a different set of genetic material.

and B, and they are made up of chromatids 1 and 2, and 3 and 4, respectively. The enzyme might nick 2 and 3 in the same place. This is followed by 'crossing over' in which the bits of chromatin marked out by the nicks in 2 and 3 swop places. So now we have a segment of chromatid 2 in chromatid 3 and vice versa. The net result is that chromosome A has acquired a bit of chromosome B, and chromosome B has a bit of A. This shuffling process, known as recombination, happens at random.

The resulting chromosomes, which are known as recombinants, are a mosaic of the two that went into meiosis. Cell division follows, and each chromosome of the pair now finds itself in a new cell. This divides again, and the chromatids of each chromosome peel apart in the process. The end result is four cells, each with a single chromosome, composed of genetic material shuffled during re-combination.

The consequence of all this is that the further apart two genes are on a chromosome, the more likely they are to part company during meiosis and the less likely it is that those two genes will be inherited together. Genes that have a one per cent chance of being parted by crossing over are said to be one centimorgan (cM) apart – a distance that turns out to be (very roughly) one million base pairs.

Linkage maps can be constructed from the results of simple genetics experiments – like Morgan's – or by direct DNA analysis of samples from families. A linkage map will tell us, for example, that the genes for enzymes A, B and C occur in the order B-C-A. It will not, however, tell us what chromosomes they are on or, if this is already known, what part of the chromosome they occupy. The physical map is needed to locate genes within the genome. The relationship between the linkage map and the physical map is like that between a register of voters and a street map. The electoral register gives the order of the houses in a street, but the street map actually locates the addresses relative to one another and to local landmarks such as a park or library.

Projects mapping organisms' genomes are time consuming and expensive, but the scientists working in this area are convinced that the genome business will have a powerful and positive impact on medicine, agriculture and basic biology. Currently the genomes of the following creatures are the subject of mapping projects: *E. coli,* yeast, *Drosophila*, the puffer fish, *Arabidopsis thalania* (land cress), wheat, apple, the pig, *Caenorhabditis elegans* (a small worm), a number of parasitic protozoa and of course *Homo sapiens*.

Already chromosomes III, VIII and XI of yeast have been completely sequenced. The report on chromosome XI appeared in the journal *Nature* in June 1994 and involved a team of 108 scientists working in 35 laboratories all over Europe. Had the journal's editor permitted publication of the full 66 448 bases of the sequence it would have filled the whole of that issue, taking up the 100+ pages of editorial, correspondence, adverts and research papers. Instead the authors confined themselves to a map of the chromosomes, flagging up the 331 genes – of which 72 per cent were previously unknown – secure in the knowledge that the full results were available to any *Nature* reader via a powerful computer database.

The yeast team had already made extensive use of this database, extracting the information that would make sense of their results and put them into context. First of all they used it to identify the open reading frames (see p. 35) in their sequence and discovered that 72 per cent of the chromosome code for protein – making it far more compact than those of more complex organisms. Then they looked at the 238 new genes they had discovered and found that 130 had sequences that resembled those of genes from other species. One of these turns out to be the gene that is defective in the rare human skin disease xeroderma pigmentosum. What the function of the yeast equivalent could be is a matter for further speculation – and experiment.

Analyses like this are turning up all kinds of genes in many different species. For instance, the *ubx* genes in the fruit fly are involved in the development of its body plan and closely resemble a group of genes that control the development of both the human and mouse nervous systems. Genes involved in human cancer have turned out to have close relatives in the worm, where their function can be easily studied. What is likely to emerge over the next few years is a new and unified picture of nature, as we begin to uncover what organisms share at the genome level.

Mapping the apple genome is another European project. The plan is to build a linkage map to speed up breeding the fruit for desirable characteristics. Identification of genes that are closely linked to other genes for traits such as fruit colour, acidity, or pest resistance will enable these to be used as 'markers'. Marker genes are very useful: if two genes are closely linked and one – the marker – is easy to identify while the other codes for some desirable characteristic, then most of the time the presence of the marker means the organism has the desirable characteristic too. If tests for these markers are developed, then they can be used to pick out desirable plants at an early stage, without the need for growing the plants to maturity to see whether they have the right characteristics.

On the mammalian side, the mouse, pig and human maps are undergoing continual refinement. The pig map, like the apple map, will have obvious agricultural spin offs. The mouse and human maps will give us fundamental information about mammalian genetics, as well as leading to new approaches to the treatment of disease (see Chapters 7 and 8). The latest mouse map is a linkage map from the Massachusetts Institute of Technology, where a centre dedicated to genome mapping has been set up. This map has an average resolution of 0.35 cM. This means there are signposts – bits of known sequence, but not necessarily genes – at roughly 750 kbp intervals along the genome. The corresponding human linkage map emerged from the company Généthon, based in Paris, who made the headlines at the end of 1993 with a physical map of the entire human genome. Généthon's linkage map has a resolution of 2.9 cM – not as refined as the mouse map, but still impressive given the size of the human genome.

Even before the formal mapping projects began, some of the characteristic features of the genomic landscape had begun to emerge. Inevitably, attention has focussed on the genes, because of their obvious biological function. In eukaryotes some genes are repeated many times. For instance, newts have up to 800 copies of the histone gene. This is one way of meeting the demand for large quantities of histone, which is needed for the rapid cell division that occurs during the early development of amphibians. In contrast, birds and mammals, whose cells do not divide so fast in the embryonic stage, have between 10 and 20 copies of the histone gene.

The genes of tRNA and rRNA are also multicopy. So great is

the cell's need for rRNA that even *E. coli* has seven copies of the rRNA gene. Ribosomal and tRNA (whose function was discussed in Chapter 2) are made by copying the gene into an RNA transcript, which is then processed by enzymes to make the appropriate species of RNA. Many eukaryotes have hundreds or even thousands of copies of these genes. For instance, the egg cell of the toad *Xenopus* has 10000 copies of one rRNA gene.

Eukaryotic genomes also contain pseudogenes – coding sequences that cannot be expressed because they do not have promoters (the 'on–off' switches that were discussed in Chapter 2). These might have arisen from mRNA that has been converted back into DNA by reverse transcription, then reinserted into the genome. There is also a viral DNA in eukaryotic genomes – scars of past viral infections, which, once they have invaded the host genome, cannot be removed.

As well as genomes, pseudogenes and viral DNA, eukaryotic genomes tend to be littered with various other repeated sequences. For example, a 300 base pair stretch called the *Alu* sequence (see p. 73) is repeated no fewer than one million times in human DNA and accounts for seven per cent of the entire genome.

The *Alu* sequence is just one of many repeated sequences. Also characteristic are the small 'satellite' DNA repeats that are found at the central region of a chromosome – the centromere. Satellite DNA accounts for 10 per cent of the genome and up to 30 per cent of other eukaryotic genomes. It appears to play a crucial role in co-ordinating the movement of chromosomes during cell division. So it keeps the two chromatids together until it is time to peel them apart for transport into the new cells.

Besides the 'junk', the repeated sequences, and the multicopy genes, one factor that contributes to the massive sizes of eukaryotic genomes is the genes themselves, which are bigger than the corresponding prokaryotic genes, even though the protein products are of a similar size. In 1977, researchers in several laboratories made the simultaneous, and surprising, discovery that eukaryotic genes are split into sections. Until then it had been assumed that they consisted of a continuous coding region like their eukaryotic counterparts. But experiments with the genes for β-globin (a segment of haemoglobin, the protein which carries oxygen in the blood) and other proteins suggested otherwise. The coding sequence of the β-globin gene, for instance, was split into three parts, bridged by two non-coding regions of length 550 and 120 base pairs, respectively.

The coding regions are called exons (because they are expressed) and the non-coding regions are known as introns. If a single-stranded DNA sequence of a gene, which contains introns and exons, is allowed to mix with its complementary RNA, which contains only the exons, the following happens. The exons match up, by base pairing, but the introns cannot find a complementary sequence on the RNA and so loop out from the DNA–RNA hybrid. These loops are clearly seen under the electron microscope, demonstrating their existence.

It is now clear that the majority of genes in the more complex eukaryotes, such as humans and other mammals, are split. There is, however, no pattern to the length and number of introns. The mouse gene for the enzyme dihydrofolate reductase (DHFR), which is used to built nucleotides, is more intron than exon. The gene has 31 000 bases in its sequence but the mRNA, after removal of introns, has only 1600 bases. The gene for chicken collagen – the main protein component of chicken bones and tendons – has 50 exons, while the gene for the human blood clotting protein factor VIII has 17 exons.

The introns are copied, along with the exons, into the primary mRNA transcript (the so-called pre-RNA) and are removed, in the nucleus, before translation. While most molecular events in the cell are carried out with the aid of protein enzymes, the removal of introns is rather different. The molecular editors with the job of excising the introns are particles of RNA and protein known as 'snurps' (small nuclear ribonucleoprotein particles, and alternatively known as snRNPs). A group of snurps descends on a pre-RNA molecule, forming an assembly called a spliceosome. Individual snurps seek out the exon–intron junctions in the gene. These are marked by a GU sequence at the 'upstream' and an AG at the 'downstream' end. The team of snurps then makes a co-ordinated attack on the RNA, which ends with the looping out of the intron and the welding together of two adjacent exons.

The discovery by Tom Cech and his team in 1983 that the

protozoan *Tetrahymena* had the ability to 'self-splice' its pre-RNA, without the aid of snurps, was an important breakthrough. It pointed to the potential of RNA alone for catalytic activity, previously assumed to be the prerogative of protein enzymes. This in turn has led to new ideas about the origin of life, which are discussed in Chapter 4, and has opened up commercial possibilities for RNA enzymes (or ribozymes).

If the structure of the same gene is examined in a number of different eukaryotic organisms, it becomes apparent that, as the organism becomes more complex, the number of introns increases. While this may suggest that introns appeared late in evolution and are in some way a reflection of the complexity of an organism, the evidence seems to suggest the opposite. By comparison of sequences of genes coding for proteins that have been highly conserved throughout evolution, it appears that introns could have been part of the genome for around a billion years. It is the simpler organisms, such as bacteria and yeast, which divide quickly and have streamlined genomes, that have discarded their introns over the course of evolution. Introns greatly increase the length of eukaryotic genes. On average they probably multiply the amount of DNA required to code for proteins about ten-fold, compared to the prokaryotic genome.

At first sight it looks as if introns are just more 'junk' DNA, which has somehow managed to invade the genes themselves. In fact, the observation that exons often code for distinct structural regions in a protein suggests that this is not the case. For instance the central exon in the β-globin gene codes for the part of the protein that binds to haem, the iron-containing molecule that both gives blood its red colour and carries oxygen molecules from the lungs to the rest of the body.

By acting as spacers between exons, introns could have been the key to mixing and matching of exons to make new proteins with new functions. This exon shuffling is now thought to have been a major driving force in evolution. For example, the exon encoding a protein called epidermal growth factor (EGF), which helps to make new skin cells, is also found in proteins that make blood clot, and also in the protein that latches on to circulating cholesterol in the body, tucking it safely away in the cells. Maybe EGF is a basic building block – sometimes known as a 'domain' – of proteins, which can be used in a variety of different contexts. It is rather like creating a wardrobe, shopping for a jacket that will go with two skirts (or pairs of trousers) as well as with your favourite jumper. EGF and other protein domains are the items of a well-planned wardrobe – and they can make a variety of protein 'outfits'.

There is also ample evidence that the alternative splicing of an RNA transcript can make two or more variants of the same protein in a cell. So if the transcript had exons called A, B, C and D, you could make proteins A B and D or A B and C (or, to push the wardrobe analogy a bit further, a black jacket goes well with a red dress and black shoes or with the same red dress worn with red shoes). An actual example of alternative splicing can be in the production of antibodies. These are proteins that the immune system uses to neutralise pathogens such as viruses and bacteria and other foreign material. They can exist in two forms. The first form has a 'sticky patch' in its molecules that lets the antibody anchor itself to a cell membrane. Foreign substances, for example bacteria, are usually noticeable to the body because their surfaces are covered in protein molecules called antigens that are not found on the host. When a bacterium bumps into a cell surface covered in antibodies, the antibodies lock onto and capture the antigens. The capture of antigens changes the shape of the antibody molecule, and this change in shape switches on a mechanism that causes the cell to divide and produce more antibody – but of the second form. This is a soluble version of the antibody, which circulates round the body and deals with all the invading bacteria. It does not have the sticky patch, because it must be free to travel. It is produced by splicing the mRNA for the antibody so the 'sticky patch' exon is excluded. So if ABCD represents the first version, then BCD could be the second version, if A represents the exon for the sticky patch domain (B C D are the other domains).

Aberrant splicing can lead to problems. In some forms of the blood disease thalassaemia, one of the splice junctions – where intron and exon meet – in the β-globin gene is altered. The spliceosome mistakenly cuts out only part of the intron in the defective globin gene, leaving the rest of it to be transcribed with the neighbouring exon. But this part of the intron contains a stop codon, leading to premature termination of translation. The haemoglobin molecule produced is too short and cannot function properly.

Thalassaemia is common in Mediterranean countries. For instance, up to 20 per cent of the population in some parts of Italy carry a defective globin gene. The only treatment is regular blood transfusion. This carries its own dangers, because the body becomes overloaded with iron from transfused blood and this has a toxic effect on the heart. It means that the thalassaemia patient has to take drugs called iron chelators, which remove the iron from the system. One day thalassaemia may be treatable by gene therapy, as discussed in Chapter 8, but for the moment it remains a severe and distressing disease.

The fluidity of the genome

DNA is far from being a fixed template, subject only to the occasional copying fault (mutation) during the replication process. It can undergo some quite radical changes during the lifetime of an organism. Sections of the molecule may hop from one site in the genome to another, or more copies of a gene may be produced in response to a chemical signal from the environment.

The first glimmerings of the fluid nature of DNA came long before its structure and functions were established. Barbara McClintock of Cornell University and, later, of the famous Cold Spring Harbor Laboratory spent many years examining the inheritance of kernel and leaf colour in maize. Noting the occasional appearance of odd coloured spots and splashes she began to wonder about the mechanisms that might control the expression of the genes for colour.

She developed the idea that there were mobile genetic elements that could jump around the chromosome, and when they jumped into a gene, they disrupted its expression. When she presented this theory at a Cold Spring Harbor Symposium, however, she was met by blank stares and indifference. The audience, no doubt, was preoccupied by phage, bacterial genetics – and DNA.

Monod's work on gene control did much to revive interest in McClintock's jumping genes. By the 1970s the existence of these mobile genetic elements, called transposons, had been established in a number of organisms. Like Mendel, McClintock was ahead of her time – but at least she was finally able to enjoy the recognition of her discoveries, when she was awarded a Nobel Prize in 1983.

Transposons are DNA sequences that can replicate themselves and form into a circle, which roams the genome, alighting at any point to break into a gene. The human *Alu* sequence, which was discussed above, turns out to be a transposon. Once it has inserted itself into a gene, a transposon has a number of effects. It can close down the expression of the invaded gene, or boost the expression of a neighbouring gene. Sometimes transposons simply cause mutation – a general messing up of the coding sequences they are sitting in.

In 1991, Francis Collins and his team at the University of Michigan found that the tumour-producing disorder neuro-fibromatosis is caused by an *Alu* transposon landing in a gene that normally regulates cell growth and turning it off. The famous Victorian John Merrick – otherwise known as the 'Elephant Man' – suffered from neurofibromatosis, and was so deformed that he was paraded as a circus freak until rescued by the physician Frederick Treves. Thanks to Treves' patronage, Merrick was able to develop his interest in the arts and enjoy the friendship of high-ranking members of society – including the Princess of Wales – for the rest of his short life (the story of the Elephant Man is told in the film of the same name).

There are also transposons that carry new genetic information into a cell. One that has serious consequences for public health is the R factor, a transposon containing a collection of antibiotic resistance genes that can shuttle between different species of bacteria. The antibiotic resistance genes code for enzymes that can break down powerful drugs such as penicillin, tetracycline and streptomycin. A bacterium armed with R factor is unaffected by many of the usual antibiotics used to treat infections. This is how almost-forgotten infections such as tuberculosis and diphtheria are making their way back into hospitals and poor urban areas, despite immunisation programmes.

That genes can pass between bacteria in this way is well established. There is now increasing speculation that transposons can pass between other species too. This may account for how one species of fruit fly can acquire the characteristics of another species, without any breeding taking place. It may be that a mite that infests one species transfers transposons called P elements from this species to another, using its mouth parts as a miniature syringe.

Gene transfer between bacteria and plants and insects has also long been suspected. Gary Strobel, of Montana State University, and his team have recently discovered a fungus that grows on the Pacific yew and can produce the anti-cancer drug taxol. Taxol was first discovered as a product of the tree itself, and is one of the hottest properties on the cancer therapy scene. It has shown very promising results in the treatment of advanced ovarian, breast, head and neck tumours. The worry is that once the drug comes onto the market, demand will far outstrip supply – for the Pacific yew is in decline.

Now it may be possible to obtain taxol easily and cheaply by growing large amounts of this fungus, but no-one knows how it acquired the ability to make taxol. Strobel suggests that it could indeed have picked up the gene (or, more likely, genes) for synthesising taxol from the yew tree.

Not only can transposons jump around and between genomes, but the chromosomes themselves sometimes undergo quite spontaneous rearrangements. Typically a piece of one chromosome will swap places with a piece of another. For instance, the Philadelphia chromosome, which is found in the majority of patients with a rapidly fatal leukaemia (chronic myelogenous leukaemia) comes from chromosome rearrangement. It is formed when a chunk of chromosome 9 and a larger chunk of chromosome 22 swop over. The resulting abnormal version of chromosome 22 is called the Philadelphia chromosome. Its formation goes hand in hand with activation of a cancer gene (of which more in Chapter 8).

Finally, if there is an environmental pressure on cells (such as the presence of a drug or a toxic heavy metal) that increases their need for various proteins, the genome might respond by amplifying the number of copies of the genes that code for these proteins. For instance, the number of metallothionein genes in cells increases if they are exposed to the toxic heavy metal cadmium (see also Chapter 2 for more about metallothionein). This results in more metallothionein to mop up the heavy metal before it can harm the cells. If cells are treated with the drug methotrexate, which blocks the enzyme dihydrofolate reductase (DHFR), most of them will die because they can no longer make nucleic acids. This is why methotrexate is used as an anti-cancer therapy, because it kills tumour cells in the same way. However, a few of the cells respond to the methotrexate threat by amplifying their DHFR gene so they can boost their supplies of the enzyme under attack. If these cells are selected, and subjected to a few more rounds of treatment with the drug, some are found to have increased the copy number of the DHFR gene up to 1000.

4

Where did DNA come from?

DNA, RNA and proteins are all molecules with a history. The study of how they have changed over time has given us a new perspective on evolution, and our place in nature. At the genetic level, evolution can be summed up as the production of new genes, their inheritance, and their selection by interactions with the environment. The fluid and dynamic nature of DNA has caused cellular life to fan out from the microbes that populated the Earth nearly four billion years ago, to the rich diversity of species we have today.

Origins

Creationists – who believe that God put each species on Earth fully formed – are conveniently sidestepping one of the toughest problems in science, that of how life began. Charles Darwin developed a convincing theory of how the earliest life forms evolved into more complex organisms. But he could not say how the first organism – often called the progenote – arose.

We will probably never know the truth about the origins of life, but there is no shortage of theories. Chemistry, cosmology and geology have all provided far more fruitful and imaginative notions about how life emerged on this planet than the creationists' stereotyped theories.

Before exploring some of the scientific ideas about the origins of life – and DNA – we ought to set the scene by trying to imagine just what our planet was like in its youth. This is not the place to go into complicated cosmological theories, so we will accept that the Universe came into being around 15 billion years ago with an event called the Big Bang. It sounds like an explosion, but it was probably more like an expansion of all matter and energy now in the Universe from an incredibly hot (about 100 billion degrees Celsius) and dense state. As the Universe expanded it cooled, and matter was formed: first the light elements like hydrogen and helium, followed by the heavier ones such as iron and tin. The Earth itself was formed between 4.5 and 5 billion years ago from a cloud of dust particles surrounding our nearest star, the Sun. The composition of the Earth – then and now – is nothing like that of living things. It has a core of liquid iron and nickel surrounded by a rocky layer called the mantle. This, in turn, is covered by our land surfaces and ocean floors, known collectively as the Earth's crust. Oxygen, silicon and aluminium – in various chemical combinations called minerals – account for more than three quarters of the Earth's crust. The latter two elements feature hardly at all in living things, whose cells are made up of compounds of carbon, oxygen and hydrogen – with a sprinkling of nitrogen and phosphorus and other trace elements.

The composition of the Earth itself has changed little, although its geography has certainly altered. But the atmosphere of the young Earth was nothing like it is today. Ideas about this have changed. Originally it was assumed that life emerged into an atmosphere made of methane, ammonia, hydrogen and water vapour. Now some scientists think that there would have been nitrogen, rather than ammonia, and carbon dioxide, rather than methane. Needless to say it is impossible to know for sure but one thing is (almost) certain – there was no oxygen.

The earliest traces of life are deposits of bacterial activity called stomatolites, which have been found in Australia, and microfossils – both of which date back to about 3.5 billion years. There is also indirect evidence from 3.8 billion year old sedimentary rocks in Isua, West Greenland. These strongly suggest that liquid water, in the form of early oceans, was present at this time. Water is an absolute prerequisite for the existence of life.

So how did the simple inorganic compounds that were around nearly four billion years ago in the Earth's crust, ocean, and atmosphere manage to assemble themselves into living cells? The origin of life is usually seen in three distinct stages. First, there has to be generation of basic organic building blocks such as nucleotides and amino acids. Next comes assembly of these building blocks into functional polymers – DNA, RNA and proteins (this is generally reckoned to be the hardest thing to explain). Self-replication (nucleic acids) and catalysis (enzymes) are the chemical processes which got life going, and kept it going. But life would not have had much of a start without the functional polymers congregating behind a barrier that separated them from the surrounding environment. Otherwise they would just diffuse away from one another and nothing of interest would happen. The barrier, the primitive cell membrane, would have formed the first cells – compartments in which the chemistry of life could evolve.

American scientists Harold Urey and Stanley Miller tackled the first part of the problem in the early 1950s, just around the time when Crick and Watson were putting the finishing touches to their DNA structure. Urey and Miller's experiments were based on earlier ideas about the origin of life put forward by J. B. S. Haldane, working in Oxford, and the Russian chemist Alexander Oparin. They argued that life had emerged from a 'hot dilute soup' (in Haldane's words). Urey and Miller tried to recreate this medium, which is often called the primordial soup, by laboratory simulations of the early Earth. To do this they circulated methane, ammonia and water through a system of flasks and tubes. From time to time they bombarded this mixture with electrical sparks, which were meant to simulate the intense solar radiation that was a key feature of the young Earth.

After a few days of all this the liquid that Urey and Miller were pumping round the system contained glycine – one of the essential amino acids. So the building blocks of life could well have emerged from the raw materials available at the time. Further experiments with different 'recipes' produced bases, sugars, and other amino acids. A joke began to circulate among organic chemists: 'Mix some ammonia and methane in a tube, leave it in the sun for a week, and then shout down the tube "is there anyone there?"'

But shock waves from meteoric and cometary impacts, and intense ultraviolet light from the Sun (there was no ozone layer to absorb it because there was no free oxygen), were hardly conducive to the orderly assembly of these building blocks into nucleic acids and proteins. Even in a quiet corner of a chemistry laboratory you need sophisticated reagents and controlled chemistry to make these biopolymers from their component parts. Just mix them in a test tube and you will get nothing, or things will go too far, resulting in a nasty black tar.

So a number of intriguing hypotheses have been suggested to explain the evolution of biopolymers. One that has been promoted in recent years by John Corliss of the National Aeronautics and Space Administration (NASA) and several other scientists is that the primordial soup was formed in hydrothermal vents, rather than on the Earth's surface. These niches, which are found on the ocean floor, were a safe haven from the meteoric and ultraviolet mayhem taking place on the surface. Although the vents would be cut off from the energy of the Sun, the energy required to drive the chemical reactions needed could have been extracted from minerals on the ocean floor.

It is hard to say how far life could have progressed under these conditions, but the proposal has been lent powerful support by the discovery of bacterial communities in hydrothermal vents. These microbes belong to a group called the Archaebacteria, which have been identified by American scientist Carl Woese as having ancient origins, and are discussed in detail later in this chapter. Indeed many archaebacteria have been found in 'primitive' surroundings that resemble our ideas about the early Earth (low in oxygen, for example).

Other scientists have argued that besides protection from the violent environment of the early Earth, the building blocks of biopolymers needed help from some form of catalyst. Graham Cairns-Smith of Glasgow University argues that clay-based minerals may not be the inert substances they are commonly supposed to be. He has developed the idea of an early stage in the evolution of life based on clay, which bridges the gap between the primordial soup and the first carbon-based cell. Clay contains silicon, which is in the same chemical group as carbon – on which life is based. Minerals are crystals that replicate their regular inner arrangement of atoms as they grow. Clay minerals can 'mutate', says Cairns-Smith, by accumulating imperfections in their crystals, which are then replicated. Even simple crystals such as those of zinc or iron will acquire these imperfections. There might be a hole in the regular arrangement of metal atoms, or a place where an extra atom has crowded in. In the case of complex minerals such as Cairns-Smith's silicates, these 'mutations' could eventually produce some form of catalytic activity. He visualises the organic building blocks that are the hallmark of life as we know it as assembling on these catalytic clays. Eventually carbon-based life took over from silicon-based life, says Cairns-Smith, who has coined the term 'genetic takeover' for the transition from clay to DNA.

More recently, Cairns-Smith has joined with other colleagues in Glasgow, such as Michael Russell, in speculating that life could have emerged from bubbles of iron sulphide (a common mineral sometimes known as 'fools' gold') in alkaline springs on the ocean floor. This environment could well have had all the raw chemicals and the energy required to generate amino acids and primitive enzymes containing iron. Similar ideas have been put forward by Gunter Wachterhauser – a US patent attorney and former chemist. Indeed, Christian de Duve of Rockefeller University has suggested that simple compounds of sulphur, carbon and oxygen (called thioesters), which formed in hydrothermal vents, might have triggered some crude early form of biochemistry.

Other scientists have avoided facing up to the problem of how the molecules of life were built on Earth by saying they came from outer space! Recent space explorations have focussed attention on the ideas of Fred Hoyle and Chandra Wickramasinghe, who have developed the concept of panspermia – life being 'seeded' by microbes falling from cometary dust. Few scientists believe this, but the discovery of organic compounds such as amino acids in interstellar dusts, meteorites, and Halley's comet suggests that the Earth may have been sprinkled at least with life's building blocks – if not fully formed organisms – in the dusts of meteoric impact.

Francis Crick, in his book *Life Itself*, takes an even more extreme view (although it is hard to tell how serious he is about it). He says that if the Universe is 15 billion years old, and life here is only 4 billion years old, then it could – in the intervening 11 billion years – have evolved elsewhere. These life forms could be very advanced and have the technology to seed life on planets such as our own. Crick envisages space ships populated with microbes being targeted to the Earth around the time when life emerged. It sounds crazy, but it is not so different from the ideas now being put forward by scientists interested in colonising other planets – of which more later.

However it happened, the appearance of a self-replicating molecule and enzymes set the scene for the emergence of the first cell. But if we think merely in terms of present-day biology an immediate problem arises. Enzymes are an absolute requirement for the self-replication of nucleic acids, as we saw in Chapter 1, but enzymes are themselves made by DNA and RNA. Which came first?

A breakthrough came in the 1980s, when Tom Cech and his colleagues showed that RNA could double as an enzyme as well as a self-replicator. Indeed, it now appears that RNA can carry out all kinds of sophisticated chemistry in the test tube without any help from protein enzymes. So RNA probably pre-dated DNA and laid down much of the biochemical groundwork for the development of the first cell. One of these bits of biochemistry was to generate DNA. Scientists at Rockefeller University believe that reverse transcriptase – the enzyme that copies DNA from RNA, as discussed in Chapter 2 – goes back a long way and that a primitive version may have created the first DNA molecule (most probably inside some form of cell, although no-one knows for sure, of course).

Once there were nucleic acids and proteins around, however primitive, they would have tended to organise themselves into cells. There is an explosion of interest today in the way big molecules sometimes self-organise into droplets and sheets. This behaviour was noticed many years ago by Alexander Oparin, who demonstrated the formation of cell-like entities from the protein gelatin and gum arabic, a carbohydrate. These droplets, known as coacervates, let substances pass in and out of their membranes. If you put enzymes inside them, they will catalyse simple reactions. Eventually these first cells may have evolved into the first single-cell organism with a DNA genome. This is called the progenote and it is the ancestor of us all. It has left no trace, but with its emergence the stage was set for the story of the evolution of DNA to commence.

Evolution – the molecular story

There are two ways of looking at evolution. The traditional story emphasises the last half billion years using fossils and archaeological finds as the main landmarks. The stars of the story are dinosaurs, apes and, of course, humans. In the molecular story, most of the action takes place in the first two billion years – with microbes as the key players.

The first microbes were probably the cyanobacteria or blue-green algae. These prokaryotes were able to photosynthesise, producing oxygen as they did so. It seems likely that, from a very early stage of evolution, microbes indulged in a primitive form of sex. This was not the sex that evolved far later in animals and plants, involving the union of specialised sex cells. Rather it was exchange of genetic material between different species of bacteria. In fact, sex is a generalised phenomenon: when a virus invades an animal cell, for example, that is a form of sex, because a new combination of genes has been formed. Bacteria were, and are, incredibly 'promiscuous' compared to more complex organisms. Studies of microbial communities suggest they have access to a common gene pool. On the early Earth, this gene swopping was a powerful protective mechanism against the intense solar radiation. Ultraviolet light from the Sun is very damaging to DNA (this is why depletion of the protective ozone layer is already causing skin cancer rates to rise). The early microbe could repair ultraviolet damage to genes simply by borrowing a spare gene from a neighbouring microbe or using enzymes to make itself a fresh copy. This wholesale movement of genes from one cell to another does not happen in eukaryotes, where genes are wrapped up in chromosomes. The gene shuffling enabled the bacteria to make a great deal of evolutionary progress. In a very real sense, we are trying to mimic this progress now with genetic engineering, which, as we will see in the next chapter, is the transfer of genes between species. The difference between bacterial promiscuity and genetic engineering is that the latter is targeted and under the control of human consciousness (itself a product of evolution). Bacterial sex is blind and random, but from time to time it is adaptive.

Then, about two billion years ago, a catastrophe occurred. Oxygen began to accumulate in the atmosphere for the first time. This had been generated by photosynthesis for over one billion years before this happened, but it had always reacted with other elements to form compounds such as iron and aluminium oxides. Suddenly, there was nothing left for the oxygen to react with and it began to accumulate. You may wonder why this mattered. After all, oxygen is essential for life. That is, however, a rather human-centred view. Human brains cannot survive for more than a few minutes without oxygen, and we rely on it to burn our food to supply biochemical energy to our cells.

The problem with oxygen is that it is the second most reactive element – it will attack most kinds of molecules in a cell, including the all-important DNA and proteins. It also damages vital structures such as the cell membrane. That makes it a lethal poison to many of those microbial life forms that get their energy without using oxygen – the so-called anaerobes (those that rely on oxygen, like humans, are aerobes). As we shall see, oxygen is eventually lethal to humans too, because of the damage it does to cells.

So the appearance of oxygen in the atmosphere would have had the impact of a global catastrophe. Billions of microbes perished; but oxygen also set the scene for a great leap forward in evolution. Nucleated cells (eukaryotes) began to appear for the first time. These cells began to develop a complex structure because of a process called endosymbiosis. This is a form of co-operation between two species of microbe, which may have begun as a predator–prey relationship. Lynn Margulis of Boston University was one of the first to pioneer the theory of endosymbiosis, which was, for many years, deeply controversial because of the way it seems to run counter to the traditional Darwinian view. Put simply, ·Darwin says that competition drives evolution, and only the best adapted to the environment survive – the rest go to the wall.

In endosymbiosis, two organisms join to become one and, in a very real sense, the new organism is more than the two that formed it. At least three of the components of modern cells – chloroplasts, mitochondria, and kinetosomes – are the result of endosymbiosis (Fig. 4.1). The story goes that bacteria, such as the so-called purple bacteria, which had learned not just to tolerate oxygen but could actually use it to generate energy, invaded species that were less able to survive in an oxygen atmosphere. Over time the two worked out a mutually beneficial relationship. The first contributed the ability to use oxygen, eventually evolving into a structure called a mitochondrion, which is the main site of generation of biochemical energy from food and oxygen in the eukaryotic cell. The second microbe may – apart from its low oxygen tolerance – have been a tougher customer than the first, giving it protection from harsh environmental conditions such as high temperature, or acidic waters.

A similar scenario could have led to the formation of chloroplasts. These structures are the site of photosynthesis in plant cells. They contain pigments similar to those of the cyanobacteria. An association between cyanobacteria and other microbes would have protected the cyanobacteria while giving its protector the ability to photosynthesise.

The discovery that mitochondria and chloroplasts have their own DNA and their own membranes inside the cell in which they reside lent powerful support for the notion that their ancestors were free-living microbes. The endosymbiotic origins of structures within the cell that enable it to move are more controversial, but the tiny cell whips that propel sperm, line our lungs and filter air, and propel single-celled eukaryotes (or protists) like *Paramecium* have a common structure. All eukaryotic cells have a transportation system that is made of protein tubes called microtubules. The arrangement of microtubules in the little cell whips is remarkably similar, whatever their origin. In cross-section, the whips all have an arrangement like an old-fashioned telephone dial of nine pairs of microtubules, with a tenth pair in the centre. Further, these whips grow out of a structure called a kinetosome, which has nine triplets of microtubules in a circle. Kinetosomes are

(a)

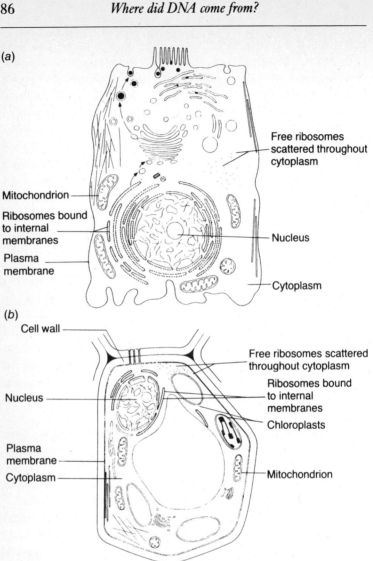

Fig. 4.1. Cell structure and endosymbiosis. (*a*) Animal cell and (*b*) plant cell. Eukaryotic cells are complex. A membrane encloses a fluid interior known as the cytoplasm. Many structures are scattered throughout the cytoplasm. Of these, the nuclear kinetosomes, the chloroplasts and the mitochondria have evolved from more primitive organisms – spirochaetes, blue-green algae and purple bacteria, respectively.

found in animal cells where they help to pull the chromosomes apart during cell division, ensuring that each new cell gets its own set of chromosomes. But reports that kinetosomes have their own DNA have not been verified.

Margulis thinks that the tiny whips and the kinetosomes come from an endosymbiotic relationship between spirochaetes and other microbes. The spirochaetes – which include *Treponema pallidum*, the bacterium that causes syphilis – are corkscrew-shaped organisms that race through their environment faster than any other microbe. Giving mobility to their endosymbiotic partners must have been a tremendous advantage: suddenly there was the ability to move off in search of new food supplies, and new possibilities for sex.

Once eukaryotes became established, other forms of DNA change became more influential in driving evolution. Mistakes made in copying DNA, which would lead to the wrong base being inserted, or extra bases or missing bases, would usually be repaired, but about one in a billion of these mutations would slip through the net. Sometimes genes, or non-coding segments of DNA, would simply be lost during cell division.

When cells become too big, they split up into a multicelled co-operative – something that began to happen about half a billion years ago (the point where we join the traditionalists with their view of evolution). In the animal kingdom the vertebrates emerged in the order fish, amphibians, reptiles, birds and mammals. Land plants came on the scene around 350 million years ago.

With the rise of animals and plants came the opportunity to mix genes in sexual reproduction, rather than in the free-for-all that is the norm for bacteria. The way genes are passed on and expressed by sexual reproduction is described by the patterns that Mendel observed in his experiments on inheritance in pea plants. He found that there were, in general, two kinds of genes: dominant acting and recessive acting. The genes are inherited in pairs, one from each parent. Each one of the pair is called an allele. If you inherit identical alleles of a gene you are homozygous for that gene (like two genes for red flowers in a runner bean). If you get two different ones, you are heterozygous for that gene.

As we have seen, some genes are linked and likely to be inherited together, but those on different chromosomes will be inherited independently. So you cannot tell if you are going to be homozygous or heterozygous for any gene (this also depends on your parents' genetic make-up). The effect each gene has on your phenotype depends on whether it is dominant acting or recessive acting.

Just to illustrate the point, here are some of the things Mendel found. When he bred a pea plant that was homozygous for smooth seeds with one that was homozygous for wrinkled seeds, all the offspring had smooth seeds. They were all heterozygotes for seed coat with one wrinkled and one smooth gene (one from each parent), but the gene for smooth coat won, because it is dominant acting. If he bred these offspring with one another, each would donate one of two kinds of gamete to the union. One would have a gene for wrinkled seeds, one for smooth seeds. So each offspring has a chance of getting either of the two gametes from either parent. That gives us a one in four chance of being homozygous for smooth seeds (a one in two chance from each parent) or for wrinkled seeds; there is also an overall one in two chance of inheriting one of each type of gene from the parents and being a heterozygote. A dominant-acting gene is always expressed. So the homozygote for smooth seed is expressed as is the heterozygote. This makes up an average of 75 per cent of all the offspring. But the wrinkled seed homozygote must give a wrinkled seed phenotype, because the dominant smooth coat gene is not present. This wrinkled phenotype makes up the remaining 25 per cent of the offspring of this mating.

So far the effect of the environment on the various mechanisms of genetic change has just been hinted at. Darwin was the first to spell this out in his great theory of evolution. He saw that there was variation both among and between species. He did not know the sources of this but realised that sometimes this variation would produce members of a species that were particularly well suited to their environment. For instance, we can imagine a family of mice with differing coat colours. If they live in woodland, the ones whose coat colour blends in best with the habitat will do best against predators. Obviously this is not the only factor influencing survival, but to make the point we will assume it is. These mice will survive longest and have the most offspring in the time made available to them for breeding. So the next generation will have more mice of the colour that blends in with woodland. Obviously no environment is static – the next year someone could cut down the wood, or a new plant toxic to mice could emerge. These factors in turn will shape the mouse population in terms of Darwinian evolution.

The forces that have shaped evolution have led us to today's rich biodiversity. No-one knows how many species there are on Earth. The lower estimate is 4.4 million, the upper is 33.5 million. We do know that nearly 1.5 million have been identified, over a million of which are invertebrates. Indeed, if you are alive, and on the Earth today, you are more likely to be an insect than anything else.

Extinction is for ever, and it is nothing new. Environmental pressures of various kinds have always driven less well-adapted species down the road to oblivion. But the rate of extinction appears to be accelerating and the gloomiest of forecasts suggests that by the year 2050 half of all species alive today will be extinct. Currently several species probably become extinct every hour; these events do not occur singly, because species are interdependent. You can demonstrate this by a simple investigation. Find an oak tree. Place a blanket under its branches and, with a long pole, gently beat the branches of the tree. A whole community of insects will rain from the branches. It is estimated that an oak supports up to 500 species (not just insects and microbes, but birds and mammals too). After counting and examining the occupants of the oak, return them to their home. Now imagine that the oak becomes extinct. Some of the species it supports may well find a new home; others would not and so would become extinct too.

Half the world's species make their home in the topical rainforest that forms a band around the equator. Satellite measurements confirm that only half of the original rainforest is left standing. Destruction of the forest has been for farming, road building, and logging of timber. Of the world's wetlands – another rich source of biodiversity – only just over half have escaped drainage and development. Overhunting, fishing, and pollution also contribute towards extinction, and closer to home urban development and road building destroy precious habitats.

At the DNA level, extinction is the loss of unique genes. Why should we care, particularly about species that were never known to science (what you never knew about you cannot miss)? Now that we hold the power of evolution in our own hands – via genetic engineering – perhaps we can shape future evolution better to suit our needs. This may well be true, but genetic engineering *depends* on knowing as much as possible about as many genes as possible. We cannot – at present – just invent gene sequences for genetic engineering. We have to find them in nature. Plants and microbes have provided us with these genes for thousands of years. So it is the worst possible kind of investment to wipe out our genetic resources.

It is tempting to think we can just take the genes, store them somewhere – perhaps even reducing them to a sequence in a computer database – and forget about the whole organism. Many scientists have now shown, however, that ecosystems with a broad biodiversity are more robust than those that are poor in species. In model experiments with ecosystems with varying numbers of species, those with the most species always responded best to various kinds of simulated climate change.

However, gene banks are being established, alongside various political and economic initiatives aimed at encouraging sustainable growth. Some take the form of nature reserves or zoos where the threatened species are looked after. Or the genes themselves could be stored in the form of germ plasm – tissue that contains the actual germ or sex cells of the species, so the whole organism can be recreated when necessary. Plant gene banks store germ plasm in the form of seeds or cuttings. The model of a gene bank was developed by the Russian biologist N. I. Vavilov in the 1920s. Vavilov also established the concept of biodiversity. Ethiopia, for instance, is a 'Valilov' centre for coffee biodiversity.

There are now gene banks in 60 countries. These include one at the International Rice Research Institute in the Philippines, which holds 60 000 species of rice – half the world's total – as well as an allotment holder's dream, the National Vegetable Research Centre at Wellesbourne in England.

There is much more to running a gene bank than just keeping the seeds in packets in locked drawers. They must be kept cold, in conditions of low humidity and from time to time they must be planted and grown on. Inevitably these conditions are not always maintained and stock might be lost forever.

And there is another aspect to gene banks that is somewhat less worthy than preserving the world's biodiversity. There is a relentless transfer of genetic material from the developing countries to the developed countries. The United Nations Development Programme (UNDP) reckons the developing countries lose at least $5 billion a year from 'gene plunder' by international drug and agricultural companies. The UNDP thinks it is time the countries of origin benefited from commercialisation of their products. More than 90 per cent of the world's biodiversity is in Africa, Asia and South America. Local knowledge and breeding skills were not being acknowledged, according to the UNDP. Genes, as they arose from a common ancestor, cannot have national boundaries or ownership. They are the common property of all humanity, but, as we shall see, it is possible to patent genes, their products and processes. The challenge is to do this so that everyone involved benefits, rather than just 'siphoning off' the genetic resources of the South to the laboratories and factories of the West.

Our place in nature

All species on Earth today share a common ancestry. Traditionally the relationships between species have been worked out by looking at their outward form and lifestyle. This classification works on a 'top down' approach and there are several systems in current use. One, the five kingdoms system, assigns living things to one of the following five big groups: monera (prokaryotes), protists (mainly single-celled eukaryotes), fungi, plants, and animals. Within each kingdom, species are subdivided again and again. Humans are animals (some people are offended by this, but we certainly do not merit a kingdom of our own – at least not on scientific grounds) and fall into a big subgroup called vertebrates. Our neighbours in this group are fish, amphibians, reptiles and birds. Because we feed our offspring with milk, have hair, live on land and fertilise our eggs internally we join rabbits, apes and mice in a 'class' called mammals. All humans are the same species, *Homo sapiens*.

Mutation of various kinds – as described above – has led to the diversity of species we know today. Sometimes these mutations, together with environmental factors such as geographical isolation, caused the formation of populations that were so different from one another that they could be classified as different species. Each continued to evolve, but from time to time, new species would again split off from the main evolutionary line. So the evolution of life becomes a vast, irregularly branched fan. All the species alive today, occupying one of the tips of this fan, can trace back their origins to the progenote. As they do so, they will meet various common ancestors at each point where the evolutionary line split.

Looking at molecules in the cell, rather than an organism's appearance and lifestyle, throws up some startling differences in the way species are related to one another. These new molecular studies have examined differences in the nucleic acids and proteins between species.

Comparison of the DNA of any two species alive today will give an indication of how long it is since the two split off from a common ancestor. For instance, humans and chimpanzees diverged around five million years ago. The differences that have accumulated between ourselves and chimpanzees are reflected in differences in our DNA. Current estimates suggest we still share around 99 per cent of our DNA with chimps. Given that five million years is not very long on the time scale of evolution, perhaps this close similarity between the human and chimpanzee genomes is not very surprising. If you look at the differences in DNA between other species, the further back they shared a common ancestor, the more different their genomes will turn out to be. In this context DNA is being used as a 'molecular clock'. The accuracy, and the quality of the information, depends on which gene is selected as timekeeper.

Genes can be classified into three big evolutionary groups, depending upon the proteins for which they code. Recent proteins are found only in animals or plants. Collagen – a major component of skin and bone – is a typical recent protein. So-called middle-aged proteins occur in eukaryotes, but not in prokaryotes. The housekeeping proteins are found in all living species; these are the ones responsible for the cell's basic functioning. Examples include triose phosphate isomerase, which is used to extract energy from glucose, and histone, the protein that DNA uses as a 'scaffold' along with other proteins (see pp. 58–60). Inevitably, the genes for these proteins accumulate mutations over time. If, however, the protein product cannot do its job because of a mutation in a critical base, then natural selection will act ruthlessly and the organism will be at a severe disadvantage. What seems to happen is that mutations in one part of the gene, which impair function, are compensated for by mutations in other regions that keep the basic function ing of the protein more or less the same. The enzyme triose phosphate isomerase does the same job in both humans and *E. coli*, but its gene sequence is only 46% similar. This makes an ideal molecular clock to explore evolutionary relationships – if it could be assumed that the mutation rate were, and had always been, the same in all organisms. Other well-conserved genes that have been used as clocks are the cytochromes, also involved in energy production, and globin, which is used in the transport of blood.

Histone is a special case. It is the most highly conserved protein in evolution. The histones of cows and peas, which diverged from one another 1.2 billion years ago, are almost identical. We can define a unit evolutionary period – the time taken for a sequence to change by 1%. For histone this is 600 million years, compared to 20 million for cytochrome *c*, 6 million for globin and 1 million for fibrinopeptide – one of the proteins involved in blood clotting.

The idea of a molecular clock was used to construct a comprehensive, and revolutionary, tree of evolution by American biologist Carl Woese in 1969. Woese's clock sequence was an RNA, rather than a gene. He looked at 16 S RNA, a component of the ribosomal RNA on which proteins are assembled and derives its name for the speed at which it sediments in a centrifuge. This 16 S RNA is found in all living things – from bacteria to humans. Woese argued that the differences in 16 S RNA between species suggested a new way of looking at the classification of life forms. Woese's kingdoms, based on the analysis of more than 400 16 S RNA sequences, are therefore based on genotypes, rather than phenotypes. His classification follows modern thinking in placing far more emphasis on bacteria than on animals and plants. I have already hinted, in Chapter 3, that from the point of view of molecular biology the really meaningful distinction between organisms is between prokaryotes and eukaryotes. Woese goes further by subdividing the bacteria into two great kingdoms: the Eubacteria and the Archaebacteria. So, according to his scheme, the progenote gave rise to three kingdoms – eubacteria, archaebacteria and eukaryotes.

The eubacteria include more familiar species such as *E. coli* and *Bacillus subtilis*, but the kingdom is extremely diverse, consisting of at least ten big subgroups. They range from photosynthetic bacteria to those that live on sulphur. Some of these shared their special biochemical abilities with eukaryotes by endosymbiotic relationships, as discussed earlier. Others merely carried on evolving independently of eukaryotes.

In recent years the spotlight has fallen, very much, on the archaebacteria. Previously this group was thought to consist of microbes that live on the edge of life in odd niches like hot springs, hydrothermal vents and very salty lakes. The archaebacteria can be divided into two big groups: the heat lovers or thermophiles are found in one group, while the other encompasses the salt-lovers or halophiles and the methane-generators or methanogens. The methanogens are found in the guts of cattle, compost heaps, stagnant lakes and in landfill sites. The methane they produce is a powerful greenhouse gas, contributing to global warming. The amount of methane in the atmosphere is increasing by one per cent every year, thanks to the activity of methanogens, which in turn are a by-product of the growth of agriculture.

The archaebacteria are often called extremophiles because of their bizarre lifestyles, and increasingly these lifestyles are being exploited for commercial reasons. For instance, an enzyme from the heat-loving *Thermus aquaticus* is now widely used in the polymerase chain reaction (see Chapter 7) which is used to amplify DNA precisely because it can stand heat. Archaebacteria that thrive on salt, pressure and alkaline, acidic or salty conditions also have commercial potential. For example, the halophiles – which live in salty water that would dehydrate and kill most other organisms – often contain purple pigments that can drive photosynthesis. These microbes are the key to photocell technology, which could turn sunlight into electrical energy. The methane generated by methanogens can be channelled off and used as a fuel, as we shall see in Chapter 11.

Surprisingly perhaps, it turns out that archaebacteria are more closely related to eukaryotes than they are to eubacteria. They also seem to be far more widely distributed than was previously supposed. For instance, they account for around 30 per cent of the microbes found in the cold surface waters of Alaska and the Antarctic. They are also unexpectedly abundant in deep waters elsewhere. According to Gary Olsen of the University of Illinois, writing about the recent discoveries in the Antarctic, our neglect of the archaebacteria has been like doing a survey of 'higher' organisms and ignoring all the animals!

It is Woese's groupings of animals, plants, and simpler eukaryotes, such as yeast, into the same kingdom – and separate from the two other great bacterial kingdoms – which has provoked the most controversy. Copernicus caused an outcry when he showed that the Earth – and, by implication, humanity – was not at the centre of the solar system (although recent surveys on the public understanding of science points to a significant pre-Copernican minority who still think that the Sun goes round the Earth!). Then Darwin pointed out the close relationship between humans and the great apes. Woese further reduces the significance of humans in nature, by placing them in a group with yeast, slime moulds, corn and frogs.

As far as 16 S RNA is concerned, as an evolutionary marker, there is more difference between Gram-positive and Gram-negative bacteria than there is between yeast and humans. The differences between 'Gram-positive' bacteria (such as *Bacillus subtilis*) and 'Gram-negative' bacteria (such as *E. coli*) can be identified under the microscope by the way in which they react with a dye called Gram's stain. Positives give a pink colour, negatives a purple colour. They also differ in their sensitivity to penicillin, an antibiotic that attacks the cell wall. Penicillin kills Gram-positive, but not Gram-negative bacteria. These differences seem trivial compared to the enormous gap between humans and yeast. Maybe Woese used the wrong clock to construct his evolutionary tree; yet, after the initial controversy, his conclusions are accepted by most biologists, and no-one has come up with a convincing alternative.

Molecular archaeology

Biochemists take enormous care of their molecules – proteins, nucleic acids, lipids and sugars – by storing samples in the deep freeze, and keeping them on ice when they are doing experiments. Letting a sample thaw out, and then refreezing it, usually ruins an experiment. Freezer failure usually means disaster. DNA, in particular, is so fragile that vigorous mixing might break up the sample into useless fragments.

It is therefore hard to believe that you could ever extract DNA from the remains of organisms that died thousands or even millions of years ago, and analyse it to get a direct read-out of the genes of the ancestors of humans and other species, but Svante Paabo, a pioneer of the emerging field of molecular archaeology now at the University of Munich, showed that authentic DNA could be extracted from 2500 year old Egyptian mummies in 1984. In the same year Russell Higuchi and Allan C. Wilson of the University of California became the first scientists to sequence the DNA of an extinct species – the quagga. This was a horse, native to southern Africa, which became extinct towards the end of the last century. Comparison with data from a current DNA sequence bank was able to show that the quagga was closer to the zebra than to other horses.

Over the course of time, DNA is degraded, as are other biomolecules such as proteins. Much of this degradation occurs shortly after death, as part of the normal decay processes. Paabo points out that tissue that dries out, such as the skin on fingers and toes, is far less prone to this sort of damage because it needs enzymes, which in turn need water so that they can work. So this type of tissue is good for extracting good quality DNA, although like all old samples it is prone to contamination by bacteria and fungi. Immense care has to be taken that this microbial DNA does not interfere with the authentic DNA from the specimen.

The main feature of old DNA is that it is fragmented into small pieces around 100 to 200 base pairs in length; by comparison, living DNA gives fragments of around 1000 base pairs for analysis. To get enough for experiment, Paabo had to clone his ancient

DNA fragments from mummies into bacteria (this genetic engineering technique will be explained in the next chapter). Bacteria find it hard to multiply such short DNA fragments. Even so, Paabo was able to show he had characteristic human *Alu* repeats in his sequences, which suggested that they came from human remains rather than bacterial or fungal contamination. He went on to practise his new techniques on an extinct ground sloth 13 000 years old and a 40 000 year old mammoth that had been preserved in the Siberian permafrost.

The polymerase chain reaction (PCR) has been of enormous help to molecular archaeology. This technique for amplifying DNA works better than bacterial cloning for this purpose because ancient DNA samples can be amplified, regardless of their size. Workers in Florida have used it to recover DNA from a 7000 year old human preserved in a peat bog in Florida. This shows a gene sequence that is not similar to those of other native Americans. Now DNA sampling of a range of native Americans is underway. Molecular archaeology suggests an exciting new way of reconstructing the population history of America.

Wilson and his team were among those who suggested that the origin of modern humans was in Africa. He looked at mitochondrial DNA (mt DNA). This is inherited solely from the mother, and is present in the ovum that is fertilised by the sperm. This DNA is far more prone to mutation than nuclear DNA because it lacks proof-reading enzymes. The African Eve – mother of modern humans – arose around 200 000 years ago, according to these studies. But these ideas are still controversial. If hominid DNA from non-African remains could be sequenced directly, then the theory could be clarified.

Molecular archaeology is pushing back further and further in time. DNA from a 17 million year old magnolia leaf has been sequenced, and the record is held by the analysis of DNA from insects trapped in amber – termites and bees in 40 million year old specimens, and 120–135 million year old weevils (conifer-eating beetles).

Perhaps the most astounding finding by the molecular archaeologist is of dinosaur DNA. Jack Horner, of Montana State University, uncovered a 65 million year old skeleton from a *Tyrannosaurus rex* – which at 47 feet (14 m) high was the largest carnivore ever to have roamed the Earth. He immediately set about extracting DNA from blood cells that he and his team found in an unfossilised part of the creature's femur. His initial analysis shows that the dinosaur DNA resembles DNA from birds. This suggests that today's birds might have come from dinosaurs. So maybe the dinosaurs did not become extinct after all – or not completely. Dinosaur experts now wonder if there may have been many more different kinds of dinosaur on Earth than they previously thought. Some may have been more like birds than reptiles. Further studies based on ancient DNA are bound to set us thinking once more about the mysteries of evolution.

Another time, another place?

Looking at how life has been synthesised from the elements on Earth and its incredible diversity and flexibility is bound to prompt speculation about whether it has arisen elsewhere in the Universe. After all, the same elements are available – albeit in different proportions. As we have seen, living things can get by without oxygen and can use a range of different chemical processes. Surely among the vastness of the Universe the conditions to support some kind of life form might emerge – or have already emerged?

To assess the likelihood of this we need to make a 'shopping list' of life's requirements. First, some kind of self-replicating and information-storing molecular system. DNA is the most powerful of these, although, as we have seen, RNA and protein can stand in for it in some circumstances. Next a control system – for example, our protein enzymes working with genes – to make co-ordinate sense of all the chemistry that drives life processes. You would not necessarily have to have nucleic acids and proteins to fulfil these functions, but it is very likely that the molecules doing this job would be based on carbon. This is because carbon can form four chemical bonds to other carbon atoms, leading to the production of straight and branched chains and rings of carbon and other atoms. There are millions of carbon-based compounds, generating the rich variety needed to fulfil biochemical functions. No other element can produce this diversity because most can link only to one, two or three other atoms. Silicon forms four bonds and produces some variety in its compounds – which is seen mainly in the chemistry of rocks. But, however hard the chemists try, they cannot get silicon to make long chains with other silicon atoms. Of course, silicon is the stuff that computers are made of, and they store information – but this is based on a completely different property of silicon and has nothing to do with chemistry! (So we will rule out computer-based life forms . . .)

So alternative life forms are likely to be based on carbon. Where should we look for them? Temperature is another vital factor. On earth the range of temperatures that can support life goes from a lower limit of a few degrees below 0 °C to just over 100 °C. Antifreeze proteins in Antarctic fish stop their blood freezing below zero in icy seas, while bacteria such as *Thermus aquaticus* have specialised enzymes that enable them to live at the temperature of boiling water. It is water that is the determining factor; it appears to be an absolute requirement of life because it provides a medium in which biochemical reactions can take place. It is a major component of the cytoplasm of the cell, and accounts for about three quarters of the composition of most organisms (more in plants such as lettuce and strawberries, and notice how spinach 'collapses' when cooking drives the water out of it). Water will work as a solvent only in the liquid state, so temperatures much below its freezing point or above its boiling point are not compatible with life. One could speculate about other solvents, such as liquid ammonia, but it would be a bizarre kind of chemistry that could support life in any medium other than water.

Besides water, and its accompanying temperature constraints, the other absolute requirement for life is an energy source. For us, that source is the Sun – a star that is near enough to provide thousands of times more energy than we actually need, but not so close that it heats up the planet to intolerable temperatures. There must be thousands of such planets occupying similar lucky positions with respect to their local star scattered throughout the Universe. Such stars do not last forever, of course. The Sun is about halfway through its estimated life of around ten billion years.

Some scientists and theologians argue that, however promising conditions elsewhere in the Universe are, life arose once only – on Earth – and that somehow the Universe has evolved with the 'purpose' that this should be so. Other scientists argue equally strongly that it is inconceivable that there is not life in abundance elsewhere. One way to settle the argument would be actually to provide evidence of this so-called extraterrestrial intelligence.

Humans have not been in space for long and we have barely managed to make the acquaintance of our neighbours in the Solar System. Of course we have increasingly powerful telescopes and a project (SETI – the Search for Extraterrestrial Intelligence) dedicated to scanning the Universe for signs of life. It is hardly surprising that the project has drawn a blank so far. Not because there is nothing to be found, but because of the enormity of the task.

An inevitable spin off from all this speculation about life elsewhere has led to the development of ideas about trying to colonise other planets ourselves. This is not science fiction with people climbing into space ships and heading for the stars, but a carefully planned re-run of evolution. One such plan, which was published in *Nature* a few years ago, argued that Mars was the best target for colonisation (terraforming, or ecopoiesis as NASA scientists now prefer to call it). The first step would be to bring its temperature up to around zero from its current chilly −53 °C. This could be done by introducing a powerful greenhouse gas into its atmosphere. The best candidate would be one of the chlorofluorocarbons (the CFCs we are so anxious to eliminate on Earth, because of the damage they do to the ozone layer). The greenhouse gas would trap more heat from the Sun. In about 100 000 years the temperature of the surface of Mars would be just about right to receive the first microbial 'seedlings', which would be sent up there to see what kind of job they would make of evolution. What would arise from this amazing experiment is of course pure speculation – unless of course a few clues came from the discovery of extraterrestrial intelligence during the 100 000 year wait!

PART II

Engineering genes

5

Genetic engineering

Knowing the structure of DNA – and how it works – has created many exciting new possibilities in the world of biotechnology (the use of biological processes to make useful products). Probably the most important technology to flow from Crick and Watson's discovery is genetic engineering.

The tools of genetic engineering allow the transfer of genes from one species to another. Because different species cannot usually breed with one another and exchange genetic material, genetic engineering opens up the prospect of creating novel species. These have the potential to widen the scope of biotechnology in ways that will have a major impact on medicine, agriculture and the environment.

For instance, the first commercial example of genetic engineering involved the transfer of the gene for human insulin to the bacterium *E. coli*. While humans and bacteria share a common ancestor, they can breed together only on the wilder shores of science fiction. With genetic engineering, a bacterium can acquire a human gene – and treat it as one of its own. In some ways there is nothing very new about this; every time you get a cold you acquire unwelcome viral genes, but the point about genetic engineering is having some control over the transfer process. Genetic engineering always creates an organism with a novel genome, although it usually only differs by one gene from its genetically unmodified counterpart. So an *E. coli* bacterium with a human insulin gene does not look remotely human, or in any other way unusual!

Sometimes the creation of a novel organism is incidental and it is the product that the organism makes which is the target of the process. Such is the case in the example described above. It is the insulin we want. No molecular biologist would dare to suggest there is any way in which *E. coli* could be improved. It has reigned supreme as the ideal experimental microbe for too long. The transgenic sheep – which are discussed in Chapter 6 – are created for similar reasons, as 'bioreactors' that produce valuable human proteins in their milk.

But genetic engineering may also focus on the organism itself. Most of this work has been done with plants to help them withstand attacks from predators or environmental stresses (this is discussed further in Chapter 10). In this chapter we consider genetic engineering as a 'production' technology.

Genetic engineering was developed in the early 1970s by Paul Berg and Herbert Boyer of Stanford University, and Stanley Cohen of the University of California at Berkeley. Patents on the technology have earned the two universities well over $20 million in royalties since they were issued in 1981. The money comes from worldwide sales of products such as human insulin and hepatitis B vaccine, which are made through genetic engineering; by 1990 these sales had soared to over $1.5 billion.

Put briefly, genetic engineering is a 'cut, paste, and copy' operation. The gene transferred is first cut out of the DNA of the organism it comes from. It is then 'pasted' into an intermediary DNA molecule called a vector, which carries it into the host organism. Here it is copied many times – or cloned – as the host organism replicates. Ideally each cell of the host adopts the new gene and expresses it as the required protein product.

The process can be illustrated by the manufacture of the enzyme chymosin, which is used to make vegetarian cheese, the first product of genetic engineering you are likely to have consumed. Cheese-making depends on the curdling action of the enzyme chymosin on milk, but chymosin is extracted from calves' stomachs, which makes cheese unacceptable to many vegetarians. Similar enzymes are found in plants, but somehow they fail to produce quite the same flavour and texture as chymosin. Genetic engineering is used to transfer calf chymosin genes to a yeast (Fig. 5.1) – so you get cheese with all the traditional qualities, knowing that the enzyme has come from a microbial, rather than an animal, source.

The first step on the road to genetically engineered chymosin is to obtain the calf chymosin gene. There are three ways of doing this. Needless to say, the genetic engineers' toolkit is full of enzymes. Among these is a set of so-called 'restriction' enzymes. In fact, the breakthrough that led to the successful development of genetic engineering was the discovery of these useful molecules by Werner Arber, Hamilton Smith and Daniel Nathans at Johns Hopkins Medical School. Restriction enzymes are found in bacteria, which use them as weapons to chop up the DNA of invading viruses. Each restriction enzyme will cut DNA at a specific sequence. They are named after their bacterium of origin, so *Eco*RI comes from *E. coli*. If a DNA molecule contains the sequence GAATTC among the millions of bases that make up its double helix, then *Eco*RI is bound to find it (along with the complementary sequence on the other strand) and snip the molecule between G and A.

Treatment of calf DNA samples (there is no need to kill a calf to get these) with a cocktail of restriction enzymes is a way of chopping up the long DNA molecule into a set of shorter bits. One of these contains the chymosin gene. This can be pulled out of the mixture using a DNA probe with a sequence complementary to a piece of the chymosin gene sequence – 20 or 30 nucleotides will be enough to track the gene down. The techniques are broadly similar to those used in genome mapping, as described in Chapter 3.

An alternative to using restriction enzymes to isolate the gene is to make the chymosin gene on a gene machine – sidestepping the need to go anywhere near a calf! A gene machine sounds glamorous, but it is just a chemical synthesiser made up of tubes, valves and pumps, which stitches nucleotides together in the right order under the direction of a computer.

Finally, calf stomach cells contain a lot of chymosin mRNA. This can be extracted from a sample of the cells and transformed into DNA by borrowing a trick from retroviruses. Just use reverse transcriptase to carry out this operation in one step.

The next part of the operation is to paste the chymosin gene into a vector. As the intended final home for this gene is a microbe, the most suitable vector for it will be a plasmid. This is a circle of DNA that occurs naturally within some microbes and will quite happily

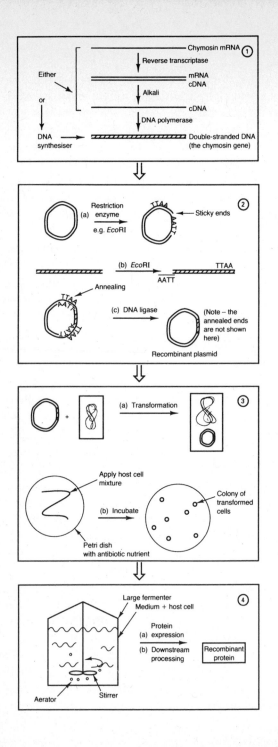

hop from one to another carrying whatever genes it happens to have within it at the time. As we saw in Chapter 4, this process dates back to the early days of microbial evolution.

To paste the chymosin gene into the plasmid, restriction enzymes once again come to the fore. Some of these cut DNA obliquely on the two strands, leaving so-called 'sticky ends', which are short, trailing sequences of single-stranded DNA. The two fragments formed by a snip of a restriction enzyme will have complementary 'sticky ends' and in theory they could 'find' one another and pair up.

Using this principle, we can cut the chymosin gene and its vector with the same restriction enzyme and encourage the sticky ends of each to seek one another out. Add another enzyme – DNA ligase – and it will weld together the sticky ends via their phosphate backbones. This forms recombinant DNA: the plasmid now contains the calf chymosin gene, and this combination does not occur in nature (although there is nothing very new or even unusual about recombinant DNA; it forms all the time during meiosis, and in communities of microbes, as we saw in Chapter 4).

Fig. 5.1. Outline of key stages in the production of genetically engineered chymosin.

1. First a stretch of DNA that codes for chymosin is produced, either from cells, or by synthesis on a 'gene machine'.
2. A plasmid vector (piece of DNA that acts as a 'shuttle') is prepared, ready for the chymosin gene to be pasted in. Base pairing between the vector and the gene produces a recombinant plasmid. This is the shuttle vehicle that will carry the gene into its bacterial host.
3. The bacterial host and the recombinant plasmid are mixed together. Some, but not all, cells will be invaded by the plasmid. The only cells of interest are those that contain a recombinant plasmid, because only these have accepted a chymosin gene. These cells are selected by the use of an antibiotic resistance 'marker'.
4. The host cells express the chymosin gene as if it were one of their own. Culturing the cells on a large scale produces large amounts of chymosin. This recombinant protein is extracted from the cells.

Now the recombinant plasmid is brought together with its host, which for the purposes of this discussion we will assume is *E. coli*. In an ideal world every bacterial cell would be invaded by at least one plasmid, and all that would be needed would be to grow the microbes as a culture, just as if we were making ethanol or penicillin. Unfortunately gene transfer by plasmid invasion is never 100 per cent efficient. Instead, the mixture of plasmid and host cell contains the following: cells with plasmids that did not take up the chymosin gene during 'pasting', cells that failed to take up a plasmid, and cells that successfully took up a recombinant plasmid. Only the third group is of any interest. So before being grown on, the cell culture has to be screened to pick out the organisms with a recombinant plasmid.

Now the antibiotic resistance properties of plasmids – discussed in Chapter 3 – can work in our favour. Often a plasmid that incorporates the genes for ampicillin resistance and tetracycline resistance is used as a vector.

Any bacteria that can grow on ampicillin must have accepted a vector – otherwise the antibiotic would kill it. The restriction enzyme cuts that open up the plasmid to receive the chymosin gene are very carefully planned to allow further selection at this stage. These cuts are made, not at random, but within the tetracycline resistance (*tet*^{R}) gene. So if the chymosin gene is pasted in, the *tet*^{R} is disrupted. This process is known as insertional inactivation.

A bacterial cell that contains a recombinant plasmid has no resistance to tetracycline, although it does, as explained above, have resistance to ampicillin. So the bacterial populations can be typed according to their resistance to antibiotics. To sort them out, the bacterial culture that results from the mixing of plasmid and host bacteria is first spread over a nutrient plate containing ampicillin. This kills off all the bacteria that do not contain a plasmid at all – leaving just those containing normal plasmids and those with the recombinant plasmids.

Obviously we cannot see individual bacteria with the naked eye – but after about 48 hours' growth they form visible blobs called colonies – each containing billions of individual bacteria. The next step is to test these ampicillin-resistant colonies by placing a sample of each on a plate containing nutrient dosed with tetracycline. Now only the bacteria containing the normal plasmid will grow. The tetracycline kills off the bacteria with the recombinant plasmid, but of course this was just a sample from the colony remaining on the original nutrient plate. Now that the bacteria to be grown on have been identified, the rest of the process resembles the traditional fermentation of substances such as penicillin (which is discussed in more detail in Chapter 9).

The use of antibiotic 'markers' in this way has led to one of the major public safety objections to the use of genetically engineered products. This argument has grown up around the production of genetically modified tomatoes, which have an extended shelf-life (the technology for producing this differs somewhat from that described above and is discussed in Chapter 9). The point is that there is a selection step for genetically modified plant cells that depends on the presence of a gene for resistance to the antibiotic kanamycin. It has been argued that eating tomatoes that contain a kanamycin-resistance gene may make consumers vulnerable to overwhelming infection.

Although kanamycin is not much used nowadays, the worry still remains and has to be confronted. It is very unlikely that the kanamycin resistance gene would not be rapidly destroyed by the acid in the stomach. DNA – for all its fragility – can be surprisingly resistant under certain conditions, but enzymes and acid will usually dispose of it fairly quickly. However, public opinion has been swayed by the problem of the antibiotic resistance gene, and threatens the commercial viability of genetically engineered foods. There may, fortunately, be a satisfactory solution: scientists in Cambridge and California working with genetically engineered tomato plants have found that it is possible to remove the marker genes that are used to create the plants. The process involves repositioning the marker gene using transposons (see p. 73) from maize. Once the marker gene has been moved, it can be deleted from the plant by conventional breeding. Alternatively, markers other than antibiotic resistance are being developed.

Careful selection of the host cell that will do the work of expressing the transgene (the transferred gene) is crucial to the success of the operation. Prokaryotic cells such as *E. coli* are easy to grow, cheap, and do a good job of producing simple proteins such as insulin. However, they cannot cope with more complicated proteins such as haemoglobin. This is because eukaryotic cells elaborate their proteins by adding short chains of sugar molecules after synthesis. These proteins are known as glycoproteins (glyco means sugar). Bacteria lack the molecular machinery to carry out this finishing touch, known as glycosylation. In eukaryotic cells there are structures called the endoplasmic reticulum and the Golgi complex, where glycosylation takes place; bacterial cells do not have these structures.

Over the past few years it has become apparent that glycosylation of eukaryotic proteins is not just for decoration – it has an important biological function. It seems, for example, that the sugars may be used for sending signals from one cell to another. Incorrect glycosylation may send the wrong type of message, leading to diseases such as arthritis. So it is sensible for eukaryotic proteins to be made in eukaryotic cells.

As far as the genetically engineered chymosin is concerned *E. coli* has been replaced by the yeast *Kluyveromyces lactis* as a host cell. The chymosin produced is 100 per cent pure and absolutely identical to the calf-extracted version in all respects.

There are many other kinds of cells that are used to express genetically engineered products: insect cells and hamster ovary cells, for instance, have been very successful. Normal human cells, however, are rather hard to culture and would never be used in a commercial operation; but human cancer cells are often used in research. For example, a culture of cells from a woman who died of cervical cancer in 1951 has been grown, divided up, and grown again many times over the last 40 years and is now used in research laboratories all over the world. These so-called HeLa cells, named after the original human donor, are not used in genetic engineering because they produce abnormally glycosylated proteins. But the cells of an African child who suffered from lymph cell cancer are used, to produce the human protein interferon, which is used, ironically, to treat some forms of cancer as well as viral infections such as herpes and hepatitis.

Nowadays, genetic engineers do not need to confine themselves to expressing genes that are found in nature. It is possible to make carefully targeted changes within a DNA sequence to make a different protein. This is called protein engineering. It might be done, say, to make an enzyme with a more rigid structure, able to withstand high temperatures. You could achieve this by strategic replacement of a couple of amino acids with the amino acid cysteine. As cysteine molecules form bridges with one another, this should toughen up the enzyme by giving it an extra supportive 'strut'. This process, which is called site-directed mutagenesis, is more likely to be successful if rational, rather than random, changes are made. For this the three-dimensional structure of the protein needs to be known – which is why protein engineers place so much importance on the technique of X-ray crystallography (see p. 21). Some of the more ambitious of the engineers are even designing proteins from scratch – so-called *de novo* proteins – to help them find out more about the basic rules of protein architecture.

6

Creating new life forms

When foreign DNA is transferred by genetic engineering to a microbe, plant or animal, a so-called transgenic organism is the result. These new life forms are created for a variety of reasons: to improve on nature, to act as 'bioreactors' that make useful products, or to act as models for understanding basic biology.

A transgenic organism usually contains just one gene from another organism within a vast sea of its own DNA. So it is hardly surprising that transgenic sheep, for example, with a gene for a human protein, do not suddenly acquire human faces (or any other noticeably human characteristics). But however reassuringly normal transgenic organisms may appear, they are somewhat different from the creatures that have emerged during the course of evolution.

How radical this difference is depends on your viewpoint. You could argue that humans have been 'interfering' in evolution since the dawn of agriculture, with the development of conventional plant and animal breeding, and genetic engineering is just a rather sophisticated breeding technology. You could point to nature's own 'genetic engineering' – the spread of antibiotic resistance, the transfer of taxol genes (see p. 74) from the yew to a fungus, and even Griffith's discovery of the transforming principle, to name but three examples. Or you may side with those who regard genetic engineering as deeply suspicious because of the way it allows the setting aside of species barriers.

Inevitably, like any new technology, the creation of transgenic animals raises a number of important issues – animal welfare, environmental and ecosystem concerns and safety. As far as business is concerned, there is also a completely new area of patenting to be explored.

Transgenic technology is truly on the move – many plants and animals have already been created. Most are still at the laboratory and field trial stage. Because animals and plants differ in the make-up of their cells and in their pattern of gene expression it has been, on the whole, easier to create transgenic plants. These are discussed in Chapter 10. Here, we will concentrate on transgenic animals.

How transgenic animals are made

The most widely used technique for creating a transgenic animal is called microinjection. This involves taking from the animal a fertilised egg, before it has divided, and literally injecting it with the DNA (corresponding to the transgene) from a tiny syringe. This must be one of the few tasks in molecular biology that has not been automated. The scientists must look down the microscope at the egg, identify the male and female nuclei, and inject just one picolitre of a DNA solution into the male nucleus, which is the bigger of the two. (This tiny amount is a millionth of the smallest volume generally handled in DNA technology – a microlitre – which is itself barely visible to the naked eye. To get an idea of scale, imagine making five billion of these injections – just to fill a teaspoon!)

The injected DNA makes its way to the chromosomes of its host, and integrates itself into some random position on the genome (in this it acts rather like a transposon, which, as we saw in Chapter 3, has the ability to jump around the chromosomes). Even though the sample injected has a tiny volume, it still contains hundreds of DNA molecules, each one a copy of the transgene. So it is quite common to find up to 200 transgenes integrated into the host genome after microinjection.

The fertilised egg begins to divide and (in mammals) is implanted into the uterus of a female animal, where it develops in the normal way. If the transgene integrates after the first cell division, then each cell in the embryo should contain at least one copy. Sometimes integration is delayed until after the first cell division. When this happens, some of the cells of the embryo will contain the transgene and some will not. The resulting animal is called a mosaic. The tortoiseshell cat, with its multi-coloured fur, is an example of a naturally occurring mosaic – some of its cells have a gene for orange fur, and others have genes for either white or black fur. It is harder to identify transgenic mosaics, because the transgene usually codes for a less obvious characteristic than coat colour. As we shall see, the possibility of creating mosaics has some bearing on the ethics of marketing transgenic animals.

Transgenic animals that result from this kind of research programmes go on to treat the transgene as one of their own. So it is inherited normally by its offspring in ways that can be calculated using the laws of classical Mendelian genetics.

Usually only one to two per cent of the injected eggs develop into transgenic animals – at least where mammals are concerned. At each stage of the process there are pitfalls. First of all, many eggs do not survive microinjection. Then around half of those that are implanted fail to develop. This has more to do with the implantation process than the fact that the embryo is transgenic. Control animals with normal embryos also fail to establish a pregnancy in about half the cases. The process resembles human *in vitro* fertilisation (IVF), which is used to treat infertile couples and is notorious for its low success rate. Finally, only a small proportion of births result in transgenic animals because the transgene may not integrate into the genome, or could be lost from the embryo during cell division.

Therefore each animal born as a result of this kind of research programme has to be screened for the presence of the transgene. Until this stage, it is not possible to find out whether the gene transfer has worked – at least, not so far. The screening is done by analysis of a blood sample from the animal, often taken by snipping off a tiny piece of its tail. The success stories are called founder animals and they are, inevitably, cherished by the researchers, who invest immense amounts of time, money and effort in their creation.

One of the first experiments to create a transgenic animal was the transfer of the gene for human growth hormone into mice. The DNA of this gene was injected into 229 embryos. These were transferred to 10 albino mice and 20 live births resulted. Of these, six mice carried the human growth hormone and went on to transmit it normally to their progeny. These mice produced the human growth hormone, were up to a third bigger than normal mice and were known as 'supermice'.

Microinjection is not the only option for creating transgenic animals. It is possible to introduce cells containing transgenes into early embryos, so that the tissues of the resulting animal contain two distinctly different cell types – one with and one without the transgene. Animals like this are called chimaeras. The advantage of creating chimaeras is that the transgene-containing cells can be selected for good gene expression before being transferred, which is quicker than waiting to see which of the animals resulting from a microinjection programme carry a transgene.

Finally cells can be removed from a mature animal for genetic modification and then returned to its body, or genes can be carried into a part of the body by a vector such as a virus. These last two techniques are the basis of human gene therapy, which is discussed at greater length in Chapter 8.

Fishy tales and the potential of 'pharming'

Fish lead the way in showing how the creation of transgenic animals can improve natural traits. It is possible to insert any gene into most common food fish, and with a higher success rate than in mammals. This is because fish eggs are fertilised, and develop, outside the body. This sidesteps the trickiness of having to remove fertilised eggs, manipulate them, then transfer them back into the body, which is necessary with animals that favour internal fertilisation.

The increasing global population puts pressure on all food producers to provide more to eat, and making fish grow faster is one way of responding to this demand. Another is to extend the ranges of food plants and animals. Since 1985, fish biotechnologists have been focussing on increasing growth rates and disease resistance as well as extending the ecological range of certain fish. For instance, Thomas Chen and his team at the University of Maryland have inserted the rainbow trout growth hormone gene into catfish and common carp embryos. They found that at least 30 per cent of these developed into fish that grew more than 60 per cent faster than their non-transgenic siblings.

Another experiment led to the creation of a transgenic version of the Atlantic salmon that can live in colder waters. The fish has been given a gene for an 'antifreeze' protein that comes from the winter flounder. This substance stops fish blood from freezing at very cold temperatures and is common among fish that live in the Antarctic.

The prospect of genetically engineering cattle for food use is likely to take many years to materialise. However, there is now genetically engineered bovine somatotrophin (BST), a hormone that boosts milk yields in cows by up to 20 per cent. BST is manufactured in *E. coli* strains that have accepted a bovine BST gene. Used wisely, BST might well make a significant contribution to increasing the efficiency of dairy farming.

In the near future, farm animals – cows, sheep, goats and even rabbits – are more likely to be used as 'bioreactors', living factories that produce therapeutic human proteins in their milk. This new technology is often called 'pharming'. Experiments with mice in the 1980s suggested that this alternative to microbial fermentation might be commercially viable and the latest figures suggest that the savings could be as much as 50 per cent (in other words, it could cost half as much to make a protein such as insulin by 'pharming' compared to making it in microbes).

The first product of pharming is likely to be the protein α-1-antitrypsin (AAT), which can be produced in quantity in the milk of transgenic sheep. AAT is an enzyme inhibitor; that is, it blocks the action of an enzyme. Enzyme inhibitors can be lethal poisons or life-savers, depending on the context. For example, the enzyme inhibitor diisopropyl phosphofluoridate (DIPF) acts as a nerve gas because it blocks an enzyme that is vital for the transmission of nerve impulses. But the action of other enzymes must be controlled by inhibitors. Blood clotting enzymes, for example, should act only to control blood loss after an injury. Once a clot has formed, inhibitors come into play to stop further clotting, which would otherwise cut off the blood supply to vital organs.

AAT, as its full name suggests, inhibits the enzyme trypsin. Its main role in the body, however, is to inhibit the related enzyme elastase. Elastase and trypsin are both proteases produced by the pancreas and are used to digest proteins in food. White blood cells called neutrophils also produce elastase as a chemical weapon against invading bacteria. Unrestrained elastase action can be dangerous, as this powerful protease starts to digest vital proteins in body tissue. AAT stops this happening.

Inherited AAT deficiency is one of the commonest genetic disorders in the West, affecting thousands of people in the USA and in Europe. The resulting destruction of lung tissue leads to the clinical condition of emphysema, in which patients have difficulty in breathing in enough air because of the loss of lung function. For people with inherited AAT deficiency, smoking is especially dangerous because cigarette smoke chemically attacks one of the amino acids in AAT that is vital in its blocking of elastase action. For a person with a normal amount of AAT in their body this may not matter too much. For the AAT-deficient it is a disaster.

AAT therapy is now being tested as a treatment for emphysema in AAT deficiency and appears to be effective, especially if it is given to people before symptoms appear. It might also be possible to use AAT for other lung disorders such as inflammation and cystic fibrosis. However, although the protein can be obtained from human plasma, the amounts (two grams per litre) fall far short of the anticipated demand.

Transgenic sheep have been created by microinjecting fertilised eggs with the human AAT gene linked to a promoter for the sheep β-lactoglobulin gene (BLG). The promoter is important: it directs expression of the AAT gene to the sheep's mammary gland so the protein is produced in the milk, from which it can easily be extracted. If the gene were expressed in some other part of the body, the sheep might have to be sacrificed to recover the protein.

This approach works. Scientists in Edinburgh have managed to get 35 grams per litre of AAT in the milk of their transgenic sheep – but hardly any in the rest of the sheeps' bodies. It is easy to purify AAT from milk and calculations suggest that a flock of 1000 sheep would be able to satisfy world demand for this protein. Other proteins that are being produced in this way include anti-thrombin III, factor VIII and factor IX, all of which can be used in treating disorders of blood clotting.

Of course, these proteins can be made by fermentation, but in smaller quantities. Pharming will score when larger amounts are needed. The transgenic animals also have the advantage of being biochemically more similar to humans than the traditional fermentation organisms such as yeast and bacteria. So they will add those finishing touches known as post-translational modification to the protein (such as attaching sugar molecules to make it into a glycoprotein, as discussed in Chapter 5) so that it has full biological activity.

The transgenic animal is both host cell and fermentation vessel. So there is no need to build a special factory to house the fermentation. Furthermore, unlike fermentation vessels, the 'bioreactor' can reproduce itself. However, it is harder to construct. Years of research and development go into the production of just one transgenic animal, with no certainty of success at the end of the project. Once a founder animal has been created, then breeding and establishment of the flock consume more time and resources. It has therefore been difficult for pharming pioneers to get funding for their endeavours.

Because development of the pharming flock takes so many years, the process is not likely to compete favourably with the traditional pharmaceutical industry, which tries out thousands of different new drugs every year and is continually changing its range of products. Pharming is most suitable for so-called replacement proteins such as AAT and factor VIII, which can be used to treat people who are deficient in them. It is possible that new clinical uses could be found for these proteins, or even that new types of milk could be produced with added nutritional or therapeutic value.

Another human protein that has attracted intense research effort – and can be made either by pharming or by fermentation – is haemoglobin. The safety of the supply of human blood has come under increasing scrutiny in recent years, with thousands being infected with viruses such as HIV and hepatitis after transfusions. So there has been a great deal of interest in the possibility of making human haemoglobin to be used as a cell-free blood substitute.

Haemoglobin is an iron-containing protein found in red blood cells and carries oxygen from the lungs to the rest of the body. It has a longer shelf life than blood and does not need to be refrigerated. Of course it is also free from the risk of blood-borne viral infections.

Haemoglobin can be produced by yeast and *E. coli*. It can also be produced by transgenic pigs that contain a human haemoglobin gene. It is made in the pig's blood, along with its own haemoglobin. The best experiments to date have produced pigs whose blood contains 50 per cent human and 50 per cent pig haemoglobin – with no apparent ill effect on the animal's welfare. Obviously the human protein has to be separated from the pig haemoglobin and all the other pig blood proteins. The pig can either be killed for its blood or used as a blood donor, with volumes of blood being removed for processing over a period of time.

However, cell-free human haemoglobin – whether produced by transgenic pigs or by microbes – does not seem to act in the same way as haemoglobin in blood. The haemoglobin molecule consists of four 'subunits' linked together. In animal tests on the recombinant product, the subunits came apart and clogged the filtration network of the kidneys, leading to severe side-effects. However, problems like this can be overcome by protein engineering.

To stop recombinant haemoglobin dissociating into subunits, scientists at the Laboratory of Molecular Biology (LMB) in Cambridge engineered the haemoglobin gene so that an extra amino acid would be added between the subunits. This acts as a bridge that holds the subunits together.

Encouraged by their success, the LMB team decided to introduce some further improvements to the human haemoglobin molecule by increasing its oxygen affinity. Haemoglobin consists of a protein called globin, linked to a smaller molecule called haem. Haem contains an iron atom, surrounded by an arrangement of carbon, nitrogen, and hydrogen atoms. Iron is the element that binds oxygen when a car body (made of steel, an alloy of iron and carbon), for example, is exposed to air and moisture.

The iron in haemoglobin also has a great affinity for oxygen.

When you breathe in, oxygen in the air moves down the lungs in passages of decreasing diameter until it reaches tiny air sacs called alveoli. From the alveoli the oxygen diffuses into a network of narrow blood vessels called capillaries. The oxygen is then bound by the iron in the haemoglobin of red blood cells. To allow the oxygen in, the haemoglobin molecule changes shape slightly in a sort of 'breathing' motion of the molecule itself. At the same time, the colour of the haemoglobin changes – to a bright scarlet colour. Blood vessels that carry oxygen to the tissues are known as arteries, whereas veins are the vessels that carry waste carbon dioxide from the tissues back to the lungs. As people trained in first aid know, venous bleeding can be distinguished from the much more dangerous arterial bleeding by the difference in the colour of the blood.

When the oxygen reaches the tissues, it is released from the haemoglobin following another change in the shape of the molecule. But only 30 per cent of the oxygen carried by human haemoglobin is actually released at this stage, the rest being returned, unchanged, to the lungs, where it is breathed out. This is not the case in all species. Crocodiles, for instance, can stay underwater for an hour without breathing – because their haemoglobin is so efficient at oxygenating their tissues. They take advantage of this species difference by drawing their prey beneath the water with them and waiting for them to drown.

A comparison of crocodile and human haemoglobin suggested that a slight variation in the chemical composition of the two proteins was responsible for the dramatic difference in oxygen affinity. The variation was narrowed down to just a few amino acids in each protein that differed. So the LMB team decided to redesign the gene for human haemoglobin so that it had codons for crocodile haemoglobin at the crucial places in the sequence. The resulting protein is mostly human, but with a bit of crocodile just where it matters.

This protein is now under trial in human patients. It is being produced by bacteria in fermenters and it could well be the first recombinant haemoglobin blood substitute on the market. A selling point will be that smaller doses will be needed, because the product is more able to release oxygen to the tissues. The product may well compete favourably with haemoglobin made in pigs. Sensing this, the US firm involved in the latter has pulled out of haemoglobin and is now developing a radically different use for its transgenic animals – as organ donors for human transplant surgery.

Transplantation and xenografts

The first human heart transplant took place in 1967 during a six hour operation at Groote Schuur Hospital in Cape Town. The recipient, a 56 year old grocer, died 18 days later. Although it was a surgical breakthrough, it remained – along with transplantation of other organs such as kidney and liver – a high risk operation for many years. The main reason was that the immune system rejects any material that is identified as foreign, such as a donor organ.

The introduction of the drug cyclosporin in 1980, however, dramatically improved the prospects for transplant patients. Cylcosporin suppresses the immune system response to the new organ, giving the latter a chance to settle down in the body. Now transplant surgery of major organs, including heart lung, kidney and liver has become almost routine and has given life to thousands of people.

The use of immune-suppressing drugs like cyclosporin, however, has its own penalties. Transplant patients are more likely than normal to suffer from various cancers such as leukaemia – a finding that lends strong support to the theory that the immune system patrols the body, eliminating rogue cells that could develop into a tumour. Rather than leave the immune system dangerously oversuppressed and the patient vulnerable to life-threatening infections, many surgeons tend to scale down the dose of cyclosporin where possible. This means that there is always a risk that the organ will be rejected; hence much research is being devoted to new drugs and treatments to overcome this problem.

The more successful organ transplantation programmes become, the greater the demand for organs. In many countries a chronic shortage has developed, leaving many people to die while on the waiting list for transplant surgery. There are several ways in which the shortage of donor organs could be eased. A shift to an 'opting out' rather than 'opting in' on the part of donors might improve the supply in places such as the UK. This would mean that a willingness to donate organs would be assumed, unless the potential donor had specifically stated otherwise. Or organs could be taken from a wider range of donors. Currently only half of the suitable organs are actually removed and offered for donation. Organs are usually taken (with permission, of course) from people admitted to accident and emergency departments. If they stop breathing, they are put on an artificial ventilator and transferred to an intensive care unit (ICU). Once declared dead in the ICU they become candidates for organ donation. The pool of such candidates could be increased by so-called elective ventilation, in which patients dying from conditions such as brain haemorrhage on general or geriatric wards could be transferred to the ICU. Here they would be ventilated – not for their own benefit, but to stop their organs from deteriorating so they can be made available for transplantation should they be declared dead. In the few hospitals where elective ventilation has been tried, the supply of organs for transplantation has doubled. However, elective ventilation raises serious ethical and practical problems.

Yet another option is to increase the use of living donors. This is widespread in Japan where removal of organs from dead bodies for transplantation is culturally unacceptable. However, this transfer of organs from a living donor to a live recipient carries a risk for the donor, and is only applicable to kidney, parts of the liver and lung, and also, surprisingly, to the heart. In this last operation, called the 'domino' technique, the donor is someone with lung disease whose heart is healthy. During the operation they receive a new heart and lungs, so a dead donor is still needed. In some countries, the use of living donors is encouraged by payment for the donated organs – while in China, where the death penalty is commonly used, there is a flourishing trade in the organs of executed criminals.

With humans proving such an unreliable source of donor organs for their own species, maybe it is time to look elsewhere – to artificial organs maybe, or to other animals for their organs. The use of artificial 'spare parts' has a very long history, from wooden legs to hip replacements that simulate genuine bone. The heart is a pump, the kidney is a filter, so using engineering principles, and the latest in materials technology, it is possible to envisage the major organs – one day – being replaced by artificial devices. However, you need only look at the amount of hardware needed to make a kidney machine, its expense and inconvenience to realise there is quite some way to go before this ideal is realised.

As long ago as the 1960s, transplant surgeons began to look into the possibility of using animals as donors of organs to humans. This type of transplant is called a xenograft, while a human-to-human transplant is called an allograft (xeno meaning foreign; allo meaning self). The early xenografts involved the donation of kidneys from chimpanzees and baboons to humans. They were not very successful – although one kidney transplanted from chimp to human lasted for nine months. It is also difficult to gain public acceptance for such experimentation with primates – partly because many of them are endangered species, and partly because they are biologically so close to humans. Since the Animals Act (1986), discussed later in this chapter, came into effect, any research involving primates in Britain is subject to special restrictions. In recent years, xenografting has been resumed in experiments that transferred baboon livers to humans – again with little success. In xenografts where the donor and recipient species are closely related like this, blood flows through the new organ for a few days before the body rejects it.

However, xenografts may still have an important role to play in the transplantation game. Transplantation surgeon John Wallwork and Cambridge scientist David White have created a litter of transgenic pigs whose hearts can be harvested for human transplantation. Pigs were chosen because they are around the same weight as humans, and their hearts are similar in many ways. It may also be that pig donors are more acceptable to the public than primate donors.

If human blood circulates through a pig heart the organ is destroyed within minutes by a mechanism called hyperacute rejection, which is too fast and too violent to be neutralised by immuno-suppressing drugs. This is only to be expected – because humans and pigs are more distantly related than humans and primates.

Wallwork and White examined the mechanism of hyperacute rejection and used it as the basis for designing their transgenic pigs. What happens is that antibodies from the human immune system activate a set of proteins that form a system called complement. These proteins work together in a ferocious attack on the foreign organ. They punch holes in it, mobilise blood clotting agents to cut off its blood supply and reduce it to a mass of blackened tissue.

The importance of complement in hyperacute rejection was suggested by experiments comparing the survival times of transplants in animals with, and without, normal complement systems. Chicken hearts transplanted into rodents with normal complement systems, such as rats, mice and guinea pigs, are rejected in a matter of minutes. However, if the hearts are transplanted into mutant guinea pigs born without a complement system, the animals survive for three days.

The human complement system will not attack an organ from a human donor. This is because the cells of the donor have several protein tags on their surface that act like white flags, keeping the peace. Wallwork and White's idea was to take one of the 'white flag' proteins, known as decay-accelerating factor (DAF) and create pigs that expressed DAF on the surface of their cells. This way, they reasoned, a pig heart can be smuggled into a human body because the presence of DAF lets it masquerade as a human heart. Accordingly Wallwork and White injected fertilised pig eggs with the human DAF gene and the eventual result was Astrid, the first transgenic pig with a 'human' heart. Born in December 1992, Astrid now heads a family of over 200 transgenic pigs. Wallwork and White are now looking at what happens when human blood is pumped through hearts taken from these transgenic pigs. The results are promising and the team hopes the first human trials of the hearts will begin around 1998.

Mouse models

The creation of transgenic animals with defective genes is the latest development in a long tradition of experimenting on animals for the sake of improving human (and animal) health. These so-called 'animal models' are used to unravel the mysteries of difficult areas of biology, such as the brain, the immune system and the development of embryos. They will also be useful for testing new treatments for diseases with a genetic basis. For example, creation of such transgenic animal models is regarded as a prerequisite for testing out gene therapies aimed at diseases such as sickle cell anaemia, haemophilia and Alzheimer's disease.

It may seem rather strange to create rabbits with Alzheimer's disease, or cats with haemophilia, because these diseases do not normally occur in these animals. However, the various genome mapping projects are confirming the remarkable similarities of species at the DNA level – and this is the rationale for using animal models. Our closest neighbour is the chimpanzee, whose DNA differs from ours by only about one per cent. Logically then, we should use chimpanzees and other primates for our animal models. In fact it is precisely this close genetic similarity between primates and humans that makes experiments on primates unacceptable to most people. Added to this, many are endangered species and they are expensive to keep and difficult to work on. So only a handful of primates are used for scientific experiments every year.

Most animal experiments are in fact carried out on mice. The mouse and human genome share an estimated 99% of their DNA (and, as we saw in Chapter 3, genome mappers have a more or less equal interest in both species). So, looked at from this angle, experiments done on transgenic mice have direct relevance to human health.

Creating transgenic mouse models differs from making transgenic animals to produce human proteins. The technology, which is known as targeted gene replacement, has been developed by Mario Capecchi and his team at the University of Utah over the last 15 years. Initially his techniques were thought to be too far-fetched to attract any funding and he had to gamble on its success by diverting money from another project. Now there are over 250 strains of mice with different gene defects – all created by targeted gene replacement and already revealing exciting new insights into the immune system, embryonic development and cancer.

Targeted gene replacement is a mixture of test tube genetic engineering and conventional breeding. It takes over a year to get the defective gene from the test tube into a mouse that can be used in experiments. The technique relies upon the use of two different strains of mouse to track the fat of the defective gene (Fig. 6.1). One strain has brown fur, the other black. The brown mice have a gene called the agouti gene, which always gives a brown coat when it is present in the genome. Black mice do not have this gene.

In the first stage of targeted gene replacement, the desired mutated gene is inserted into cells from a brown mouse embryo called embryonic stem (ES) cells. ES cells are said to be pluripotent; that is, they can develop into any type of cell as the embryo matures. The gene insertion is done using the genetic engineering techniques described in Chapter 5. When the vector with the mutated gene is inserted into ES cells, three things can happen. It may fail to integrate into the chromosomes at all, or it may integrate randomly. Alternatively – and this is the most desired yet least likely outcome – it might undergo a process called homologous recombination, which was discovered by Capecchi himself in the late 1970s. In homologous recombination, the new gene wanders up and down the chromosome until it finds its normal counterpart. It then lines up next to it, and the two genes swap places. The normal gene is thrown out of the chromosome and the defective one takes its place. This amazing event – which opens the door to much more precise control over genetic manipulation – occurs only in one in a million treated cells. No wonder the grant awarding committees were sceptical!

Antibiotic resistance genes are used to pick out the precious cells that have the defective gene in the right place on the chromosome from the rest, using techniques similar to those of other genetic engineering experiments. Next a black mouse is brought on the scene, to donate an embryo into which the ES cells are injected. The treated embryo is then inserted into the womb of a third mouse, whose colour is immaterial, and left to develop.

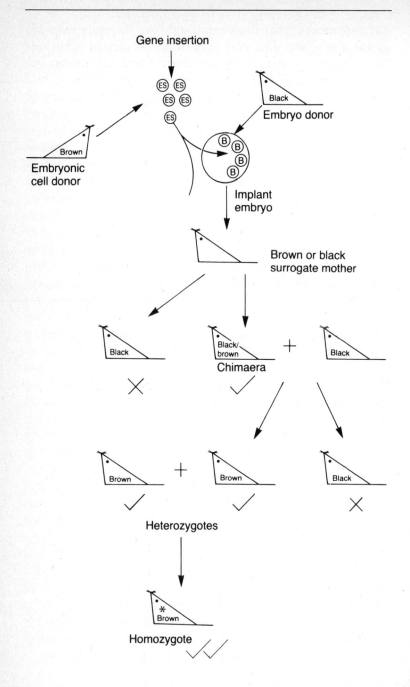

Two sorts of mouse can be born from this process. If the transferred ES cells do not survive the experiment, the baby mouse has black fur. However, if the ES cells are alive and flourishing, the mouse is a chimaera with black and brown fur, reflecting the two types of cell in its tissues – the ES cells (from the first mouse) and the cells of the embryo (from the second mouse).

Now breeding experiments can start. From a chimaera, a mouse with defective genes in all its cells must be produced. The germ cells of this chimaera may come from the brown mouse (the one with the defective genes) or from the black mouse (with normal genes). The chimaeras are mated with black mice. From this, either brown or black offspring are obtained, depending on what kind of germ cell the chimaeras contributed to the union. These offspring are no longer chimeras, because they have been bred in the normal way. The brown mice are the ones that have inherited the defective gene, but it will be in only one of its paired chromosomes (the other will be normal, because it came from the other parent, which was a normal mouse).

If these mice were used in experiments they would not display fully the physical characteristics of the defective gene. The presence of the normal gene will protect the animal from this. Such mice, however, are carriers of the defective gene. If they are bred with one another there is a chance (actually a one in four chance

Fig. 6.1. Targeted gene replacement. This process creates a mouse with a specified gene defect. First, cells with this gene defect (ES cells from a brown mouse) are injected into an embryo from a black mouse. This embryo develops in the womb of a surrogate mother. If the ES cells survive, the baby mouse has black and brown mouse cells and will have a mixture of black and brown fur. This chimaera is bred with a black mouse. Offspring with the gene defect reveal themselves by having brown fur, because of the origin of the cells with the gene defect. Further breeding is necessary to create a mouse bearing two genes with the defect. These mice can be used as a model for studying this defect. Mice with only one defective gene may be protected from its effects by the other, normal, gene and are of limited use. ES, embryonic stem cells from a brown mouse and carriers of the inserted gene; B, cells from a black mouse.

according to Mendelian genetics) of producing a mouse that has two copies of the defective gene, one on each chromosome of the relevant pair.

Capecchi has used targeted gene replacement to look at the role of individual members in a family of genes called *Hox*. Experiments on fruit flies showed that *Hox* is important in an orderly unfolding of the organism's body plan – directing limbs and organs to the right place, and ensuring they are the right shape. Fruit flies have eight *Hox* genes, whereas mice and humans have 38. Capecchi found that inserting a defective form of one of these, the *HoxA*-3 gene, into mice causes multiple malformations of the heart and its blood vessels, the thyroid, head and throat. But all these organs derive from a particular group of cells in the embryo. So it seems that the development of this section of the embryo is under the direction of *HoxA*-3.

Defects in genes involved in development can lead to the birth of children with malformations. These account for a significant number of childhood hospital admissions, and can lead to a whole range of problems – from shortened life expectancy to discrimination and lack of opportunity. A better understanding of the genetic basis of malformation might lead to new ways of reducing its incidence.

Animal rights and wrongs

By now you could well be feeling a little uncomfortable – if not outraged – at the implications of DNA technology for animal welfare. Widespread use of transgenic animals is probably still a few years away, and the use of animal models will always be a research tool. However, there is little doubt that these developments are going to be unacceptable to people from a number of special interest groups, as well as to the ordinary citizen.

There is a long tradition of using animals for medical research, and the practice has always been controversial. In 1986, the Animals Act was passed in the UK to regulate animal experiments. The Act, besides requiring scientists to be licensed to carry out this sort of work and providing for veterinary support, allows for the collection of statistics every year about the numbers and types of animal experiments. These show that – perhaps contrary to public belief – the number of experiments is on the decrease (largely because of the development of non-animal alternatives), most of them are carried out on rodents, and only a tiny percentage are concerned with testing toiletries and cosmetics. A similar downward trend is apparent in the Netherlands – the only other country to keep detailed figures.

The Act says that the likely benefits to humans (or, in the case of veterinary research, animals) from experiments have to be greater than the suffering they cause to the animals involved. In 1990, statistics on experiments involving breeding transgenic animals were collected under the requirements of the Act for the first time and around six per cent of experiments fell into this category.

Since then, there has been little public discussion about how the benefits/costs requirement of the Act apply to transgenic animals. Many animal welfare groups reject any work with transgenic animals out of hand. Increasingly, scientists are reluctant to even admit in public that they work with animals – let alone discuss their feelings about the subject – for fear of violent reprisals from animal rights activists. Therefore, opportunities for real debate are limited.

As far as using animals for 'pharming' is concerned, there seem to be no significant adverse effects on health and welfare. This is probably because the proteins are produced in the milk – creating few demands on the rest of the animal's biochemistry and physiology. However, in the early experiments on transgenic animals, where growth hormone was inserted into the genome to create bigger animals, some animals developed severe arthritis as a result of the treatment.

The development of animals as sources of human organs is going to be harder to justify to the general public and animal welfare groups, although its development has been allowed by the Act. So far, Astrid (the pig with a human heart) and her family seem to be in good health, but they will be sacrificed for their organs. Obviously pigs are sacrificed for their meat the whole time, so you could argue that this is nothing new. Objections to this work may arise more from the feeling that there is something undignified for the recipient in receiving a pig's heart, and that scientists are interfering with nature to an extent that is bound to backfire. However, it is hard to find many scientific inventions – particularly in the medical sphere – that do not interfere with nature to a certain extent. This, in itself, is no guarantee of failure.

As far as animal models are concerned, this work inevitably involves animal suffering. What has to be done is a careful analysis of whether the benefits really justify this work. It is far too early to say where the balance lies.

Patent protection

Patents are a form of protection for the investment of time, effort and money that is put into an invention. Although many scientific discoveries, such as the genetic code and the structure of DNA, can be seen as part of human culture – once due credit has been given to the scientists involved – funding for research would very soon dry up if more specific ownership was not allowed for some of the products that flow from such discoveries.

A patent gives an inventor a monopoly on his or her invention for a limited period of time. In the pharmaceutical industry, for instance, the patent on a drug lasts for 20 years in Europe. However, not all inventions are patentable. To qualify, the invention has to be novel, non-obvious and useful. The inventor has to supply enough detail in the patent description to allow the invention to be reproduced by others – but if someone else tries to make or use it, they have to pay royalties to the inventor.

Biotechnology has led to many inventions that are theoretically patentable, from technical processes through to genes, plants, and animals themselves. It might seem odd that you can patent living things, and for many people it goes against the dignity of nature to allow it. However, these are organisms that do not occur naturally, so they are 'inventions' as far as patenting goes.

Patent law differs around the world. As far as biotechnology is concerned, the important areas are the USA, Japan and Europe.

These differences are a constant irritant to the biotechnology industry, because they may affect the competitive edge a company has in certain markets. Both the USA and Japan allow the patenting of transgenic animals. In Europe, with the differing views of several countries to take into account, the situation is less clear and shows little sign of being resolved as time goes by.

Recent discussions aimed at formulating a new European Directive in this area suggest that some inventions might not be patentable because they would be against 'public order'. As far as transgenic animals are concerned, this means inventions where animal suffering is involved without any clear therapeutic benefit for humans or animals (which reflects the requirements of the UK Animals Act). Already animal rights groups are gearing themselves up for a campaign against patenting of transgenic animals, citing the public order clause.

The case of the first genetically modified animal to be patented highlights these issues. 'Oncomouse' is a mouse engineered by a team at Harvard University. It has an oncogene (cancer gene) that makes it more susceptible to developing tumours than do normal mice. The United States Patent Office granted a patent to Harvard for the Oncomouse in 1988 and the patent holders then licensed this to a company that makes the mice commercially available.

The mice are used for testing new drugs and other chemicals to see whether they cause cancer. Because they are so prone to tumour formation, they are more sensitive to these chemicals than other experimental animals. The patent is quite broad, covering not just mice bearing the oncogene, but other animals carrying it as well (in fact, the only animal that cannot be patented under any circumstances is a human). To protect their investment while marketing in Europe, the investors wanted a European patent as well. Otherwise rival companies would create their own oncomice and sell them without having to pay any royalties.

Initially the European patent office refused to grant the patent, citing the public order clause, being unconvinced that the animal suffering was compensated for by any use to humans in carcinogen testing. However the decision was overturned on appeal, and the patent was granted.

This is not the end of the story. Animal welfare groups – acting as a broad European coalition – are trying to have the patent withdrawn, again on the public order ground. They argue that there are other, non-animal, methods for carcinogen testing and that sales and use of Oncomouse to date do not justify the suffering caused to the mice.

Public opinion is likely to play a large role in whether future patents of this kind are granted. If investors in companies that rely on transgenic animals – for instance in making therapeutic proteins – cannot get the public on their side, then cash for research and development will dry up. Without a patent, there is no finance to be earned from royalties. Therefore biotechnology companies with a stake in the transgenic animal field are watching the patent scene with interest and concern – their futures depend on it.

Food for thought

At present transgenic animals are among the most valuable commodities of the biotechnology industry. They are far too important to be allowed into the food supply. But it is only a matter of time before transgenic cattle, poultry or fish start to make their way to the supermarket.

Recently, one firm involved in 'pharming' decided to test the waters of public opinion and asked the British Ministry of Agriculture whether their 'failed' transgenic sheep could be allowed into the food supply. The Ministry consulted a number of interest groups and found that, while few people were against genetic engineering per se, many were worried about the prospect of finding human genes in their food, or genes from animals they were forbidden to eat for religious reasons. A related possibility is the introduction of animal genes into plant foods, which would perhaps be unacceptable to some vegetarians.

Besides causing offence to people for ethical reasons, the introduction of foreign genes into food animals could cause health problems. Foreign genes, if expressed, produce foreign proteins that could provoke allergies. Allergies to peanut proteins, for instance, provoke a violent reaction that may even turn out to be fatal in a few people. It is worth noting that the flesh of the animals used in 'pharming' would contain copies of human DNA, but no human protein because the gene is expressed only in the mammary gland (so the corresponding protein occurs only in the milk). Fears that the foreign genes themselves could be harmful are probably unfounded because the human gut breaks down DNA.

Labelling foods that contain transgenes might seem to be the only option that gives the consumer real choice. Some manufacturers object to this where the nutritional value of a food produced using some form of gene technology does not differ from a version produced in the conventional way. Some shops have launched a blanket boycott of any food produced by genetic engineering. The first widely available product from gene technology is vegetarian cheddar produced using recombinant chymosin. In the UK, the Co-op is currently the only retailer that labels this product and offers an explanatory leaflet on genetic engineering. So far customer response to this initiative has been positive.

7

Genes – the human angle

Molecular biology and the DNA revolution are already having an enormous impact on medicine. There is an increasing emphasis on the role of genes in disease, along with powerful new DNA technologies for exploring an individual's genome. At the same time, the molecular approach has given important insights into some of the toughest problems in biology: development, ageing and cancer. We even have the tools to change our genes or control their activity, using gene therapy.

The basics of human genetics and the different DNA testing methods are looked at in this chapter. The new medical research findings that have emerged from molecular biology, along with the prospect of DNA-based therapies, are reviewed in Chapter 8.

Human genetics – a basic guide

The DNA molecule at the heart of each cell in the human body is like a signature, unique to each individual. There is only one exception to this rule. Identical – or monozygotic – twins have identical genomes. The reason is that they develop from a single fertilised egg that splits into two embryos sometime during the first two weeks of a pregnancy. As we shall see, research on twins is important in trying to assess the relative contributions of genes and the environment to someone's physical and mental make-up.

Within the three billion base pair sequence of the human DNA molecule there is obviously plenty of scope for variation in the ordering of the four bases. At the same time, all human DNA molecules have a broad similarity, which lets us all function as members of the same species.

Most of this variation is within the 'junk' DNA that does not code for protein (see p. 57). It is useful for identification purposes, but as far as we know, does not have an impact on our phenotype; that is, our physical identity. By contrast, variations within the genotype – the collection of genes you inherit – can lead to very obvious differences in characteristics such as facial features, eye colour and height. For a significant minority, differences as small as a single base can result in devastating disease, such as sickle cell anaemia. As is becoming increasingly apparent, more subtle variations – probably involving many genes – may be translated into susceptibility to heart disease, cancer and other common illnesses.

Sexual reproduction is the key to the genetic individuality of humans and other organisms. When the male and female germ cells (sperm and egg) unite in fertilisation each makes a DNA contribution to the new individual in the form of a set of 23 chromosomes. Each chromosome in the germ cell has undergone crossing over and recombination during meiosis, as we saw in Chapter 3. So each one contains a random selection of genes from the parent. Fertilisation leads to pairing of the chromosomes, resulting in a diploid cell. We can divide the cells of the body into two types: germ cells, which are haploid; and somatic cells (all the rest), which are diploid. If you look at somatic cells under the microscope you will see that for chromosomes 1 to 22 the chromosomes within a pair look similar. One comes from each parent. Genes that occupy the same position on each chromosome in the pair are called alleles. These may or may not be the same. So someone could have identical alleles for the globin gene, having inherited the same gene sequence from each parent, or they may have different alleles for the gene (they both code for globin but for different versions of the molecule; they may both function equally well, or one may be defective). The remaining chromosome pair are the sex chromosomes – X and X in a female, X and Y in a male. The X chromosome is bigger than the Y chromosome and the latter carries very few genes.

If germ cells were diploid, then the fertilised egg would be tetraploid; that is, each chromosome would occur as a set of four. While multiple chromosomes like this are common in plants such as wheat, in animals they are associated with birth defects and sterility. For example, in Down syndrome – one of the commonest of all chromosomal abnormalities, occurring in one in 700 live births – there is an extra copy of chromosome 21.

No-one knows why this rogue chromosome leads to the facial features and mental retardation that typify Down syndrome. Such chromosome defects are quite common, with a frequency of 20 per cent of all pregnancies, but because such embryos usually abort – typically before a woman even knows she is pregnant – the number of babies actually born with a chromosome defect is around six per 1000. These aberrations are sometimes the result of mistakes during meiosis, and that these occur is hardly surprising when you consider how complicated the process is.

The phenotype that develops from the fertilised egg depends, to a greater or lesser extent, on the selection of alleles that the parents have handed down. It is possible to see Mendel's laws operating in human inheritance, but it is not nearly as easy as in simpler organisms such as fruit flies and peas. For a start, humans do not have big enough families to observe Mendelian patterns with any consistency (a point we will return to). Also genotype does not always translate into phenotype – a phenomenon called incomplete penetrance, and last, but not least, the effect of the environment on the development of a phenotype has to be taken into account.

Work on human genetics has focussed on the inheritance of genes for disease, rather than for 'good' characteristics. This is partly because the inheritance of at least some diseases is far better understood than the inheritance of beauty or good nature (if these qualities are inherited at all). While few people object to breeding plants for big flowers, high yield of fruit or frost resistance, the prospect of equivalent experiments on humans is generally considered to be out of bounds by the general public and scientific community alike.

The burden of genetic disease

All parents hope for a healthy baby, and in around 95 per cent of cases their wish is granted. The unlucky five per cent are born with some kind of congenital disorder. Some of these – such as various forms of mental retardation – can be attributed to problems during pregnancy and childbirth, such as maternal infection. The rest – probably around two per cent – have problems that are rooted in defective genes which they have inherited from their parents.

Genes become defective by mutation, which alters DNA in some way. As we saw in Chapter 4, mutation is not always harmful: it is one of the driving forces of evolution. Given that DNA is constantly being replicated and divided between new cells in mitosis and meiosis as well as being exposed to various environmental influences, it is hardly surprising that it tends to change over time.

DNA analysis of patients with family histories of various diseases has shown that there are a number of ways in which a gene can become defective via mutation. First, the gene can simply be missing from the genome. There are several ways in which this could happen. Loops containing the genes could form and then be excised by mistake from the genome. Or there could be misalignment of gene sequences during crossing over, which could leave some haploid cells without a particular gene.

Less dramatically, the wrong base could be inserted into the DNA molecule during replication. This is called point mutation and it turns the codon in which it occurs into a different codon. The effect on the organism varies. Sometimes – thanks to the degeneracy of the genetic code – there is no effect because both normal and mutated codons code for the same amino acid. This is called a silent mutation. A missense mutation is one in which a change in codon results in a different amino acid being incorporated into the protein for which the gene codes. How serious this is for the organism depends on how crucial that particular amino acid is. For example, sickle cell anaemia, discussed later in this chapter, is caused by a missense mutation. Finally, a point mutation could turn a codon for an amino acid into a stop codon. This means transcription will stall when it reaches this point in the gene. A truncated mRNA leaves the nucleus and is turned into an abnormally short protein. It is unlikely that this will be able to fulfil its biological role because it will be too short to fold up into its correct shape. This is called a nonsense mutation.

These point mutations happen about one in every 10000 base-pairing events. The reason is that the four bases are all present in the nucleus together, and their chemical structures are not all that dissimilar. So every now and then DNA polymerase, the enzyme that directs the synthesis of new DNA, is going to insert the wrong base by mistake. If this were the true frequency of mutation, however, it is hard to imagine the consequences for living things – because the actual rate of mutation turns out to be one in a billion base-pairing events. The reason for this discrepancy is that DNA polymerase works with a team of 'proof-reading' enzymes, which correct the mistakes in its copy. At least, these enzymes have been found in bacteria and their presence in eukaryotes is assumed. To approach the incredible accuracy of DNA proof-reading, just try typing out a quarter of a million pages of text without single error!

Remember that the human genome has 3000000000 base pairs. At this mutation rate, there are about three errors every time it is copied. Inevitably there will have been a few mutations in your own DNA as you have been reading this. Most will be in your 'junk' DNA and will have no noticeable effect. In any case, most mutations that slip through the proof-reading net are dealt with by a set of 'repair' enzymes in the cell.

The mutation rate is increased by environmental stresses such as certain chemicals, radiation and some viruses. These are called mutagens and they can lead to cancers during the lifetime of the individual – as well as affecting their germ cells – if they happen to occur in genes that are important in controlling growth and division of cells, as is discussed in Chapter 8.

The most obvious genetic disorders are the single gene defects. Over 5700 of these have been recorded. Some are extremely rare. For instance, acute intermittent porphyria, which was responsible for the 'madness' of King George III, is a disorder in the metabolism of haem (the pigment that makes blood red), which occurs in one in 100000 births in northern Europeans. The commonest single gene defect in this group is cystic fibrosis (CF), which affects one person in 2000. Single gene defects probably account for around 14 births in every 1000. These figures mask the significant impact certain disorders have in some communities. For example, Tay–Sachs disease, which leads to blindness and mental retardation, is 1000 times commoner in Ashkenazi Jews (a group of Jews of German and eastern European descent) than in the rest of the population.

Singe gene defects fall into three categories. Those that occur in genes on chromosomes other than the sex chromosomes are called autosomal disorders. These are subdivided into dominant and recessive disorders. The third type are the so-called X-linked disorders; these arise from a gene defect on the X chromosome. Being clear about these categories is essential for those on the receiving end of genetic counselling because it helps them assess the chance of having an affected child.

Thousands of adults, and their families, now face the agonising prospect of discovering whether they carry a defective gene that leads to Huntington's disease (HD), a cruelly disabling brain disorder that develops typically in early middle age, when the gene is likely to have already been passed on to the patient's children. Because HD is autosomal dominant, if you have the gene you get the disease, even if the other allele is normal (it is very unlikely that two people with HD will marry). The chances of your inheriting the HD gene from the affected parent is 50 per cent – because they have one normal allele and one HD allele at that point on the chromosome. Roughly half their germ cells have a normal allele and half have the HD allele because the paired alleles are separated during meiosis.

There is no cure for HD. Before there was a test, people just had to wait to see what would happen if they had a parent with the disease. Now they can know in advance what their fate will be. Obviously there are enormous benefits (but perhaps some guilt directed towards affected siblings?) for people who turn out to be in the clear. But what about the psychological pressures on those who are not so lucky? One consolation is that now that the gene has been found, new treatments such as gene therapy can be planned. At the very least, scientists can begin to understand the underlying molecular pathology – the genes and proteins involved – in the disease. This is what happened in the case of CF where the defective gene was discovered five years before the HD gene.

In CF a defect in a protein called CFTR (cystic fibrosis transmembrane conductance regulator) leads to an imbalance in water and salt transport into and out of cells. The result is the production of thick, sticky mucus in the lungs, which is a breeding ground for infection. Even with the best of care, few CF sufferers can expect to survive past 30 years of age. However, discovering the gene has led to a far better understanding of the disease – because the gene led straight to the protein it codes for (CFTR) and the revelation of how it goes wrong in CF. So now a gene therapy programme is under way which will, it is hoped, dramatically improve the prospects of the young people who suffer from CF.

CF is autosomal recessive. So having one CF allele and one normal allele means you are only a carrier. To have the disease, a person has to have two alleles. If two carriers marry, there is a one in four chance of having a child with the disease – as each parent has a 50/50 chance of transmitting the CF gene. If a carrier marries a normal person the worst that can happen is a child being born who is a carrier – of which there is a one in two chance. What is vital to convey to people at risk is that the die is thrown afresh at each pregnancy. So if a couple, both carriers, have one affected child it does not mean that the next three will be unaffected, or just carriers. The next child has the same chance of being affected as the first. Other common autosomal recessive disorders include sickle cell anaemia, with a carrier frequency of one in three in African Blacks, and thalassaemia, which was discussed in Chapter 3.

X-linked diseases such as haemophilia and muscular dystrophy usually affect only boys, whereas girls are carriers. This is because a male has only one X chromosome; if it bears a defective gene, then he will develop the trait with which it is associated in a dominant fashion. If a girl inherits the same faulty gene, however, the trait behaves as if it were recessive, because she also has a normal X chromosome (except in the very rare situation where she has a carrier mother and an affected father, where there would be a 50 per cent chance of her inheriting two defective genes).

There is little doubt that the impact of genetic disease on the family, the health services, and the community at large is on the increase – for two reasons. First, with the development of powerful antibiotics and effective vaccines the threat of infectious diseases such as tuberculosis and smallpox has receded (although complacency on this point would be dangerous, given the emergence of AIDS and the resurgence of malaria and tuberculosis around the world). This means the relative importance of genetic disease has increased. Second, better treatment means that more people with the genetic disease survive for longer – although their life expectancy is often shorter than average. This means that many families have to care for many years for a child or children who may need repeated hospital admissions. For instance, people with inherited blood disorders such as thalassaemia and haemophilia often need repeated blood transfusions, or blood products – at great cost to the health service.

Advances in gene technology have also led to greater concentration on the genetic components in the chronic conditions that are the leading causes of ill health and mortality in the industrialised and affluent nations. These include heart disease, high blood pressure, diabetes, asthma, cancer and infection.

However, these are diseases that are also strongly affected by lifestyle and environmental factors. Family studies involving detailed study of medical records have traditionally been used to try to assess the relative contribution of genes to diseases of this kind. Twin studies, which were originated in 1975 by the English anthropologist Francis Galton, are particularly important here. There are two kinds of twins: identical twins (monozygotic) come from the same fertilised egg and so have the same genotype, whereas non-identical twins (dizygotic) come from two eggs fertilised by two separate sperm. The genetic relationship between non-identical twins is the same as between siblings who are not twins.

If twins share a particular trait they are said to be concordant for it, and discordant if they are not. In one study of identical twins, 30 per cent of them were concordant for high blood pressure. If this were a single gene disorder, concordance would be 100 per cent because the twins have the same genes (assuming complete penetrance – that is that the 'high blood pressure' genotype always leads to the development of high blood pressure). In the same study, the concordance of the non-identical twins for high blood pressure was only ten per cent. The difference between the two kinds of twins suggests a genetic component for high blood pressure because the more genes the twins share the higher are the concordance rates for the actual genes involved. Twin studies of this kind were also important in sorting out the factors behind the two main kinds of diabetes. Juvenile diabetes – the more serious form of the disease, where the patient is dependent on insulin – has a weaker genetic component than does maturity (or late-onset) diabetes.

Another way of looking at the genetic factor in disease is to investigate the DNA of people suffering from a particular illness in the hope of finding an association. This approach, where DNA is tested, tells us more than just looking at medical records. It may reveal which gene or genes are involved in the disease or, more likely, what genes or genomic features are commonly inherited with the disease (so-called linked markers).

A recent find of this kind that caused much excitement was of a mutation that occurs in people who have heart attacks when there are no obvious risk factors such as overweight, age or smoking. Most of us know someone who did all the 'right' things – exercise, low fat diet, no smoking, alcohol in moderation and so on – and died of a heart attack, often at a tragically young age. Sometimes, if you look at this person's family history, you might find a relative similarly stricken. It is possible that such people have a mutation in a gene coding for angiotensin-converting enzyme (ACE), which is important in controlling the blood pressure and the overall state of the blood vessels. A healthy heart depends on the blood vessels supplying it being in good condition. If they get narrow or blocked the blood supply could be cut off. The resulting lack of oxygen will kill the heart tissue, resulting in a heart attack, or to give it its clinical name myocardial infarction (MI). Up until the 1950s MI was rare. Now it has reached almost epidemic proportions in the West. Doing something about heart attacks has to be a high public health priority, and uncovering the genetic component is an important part of this. Following the discovery of the ACE mutation in 1991, some doctors argued that the population should be screened for this, and those found to carry the mutation given drugs that normalise their ACE function, thereby decreasing their risk of a heart attack.

It is easy to accept that heart disease, diabetes and high blood pressure might 'run in the family'. What is more surprising is the notion that there could be any connection between genes and infection. Microbes that cause colds, flu and sore throats appear to attack without discrimination. Yet there are always some people who are hit harder than others, and a few who will invariably resist the latest 'virus' that is going around the office or the classroom. Scientists working in the Gambia have recently managed to pin down a gene that makes people more susceptible to infection. They looked at children who were particularly hard hit by cerebral malaria and found they were producing abnormally high levels of a protein called tumour necrosis factor (TNF). This appears to happen because these children inherit an abnormal variant of the TNF gene, from each parent.

It now appears that most physical illnesses may have some form of genetic component. But what about the psychological angle? Are mental illness, social problems, personality, intelligence and even sexual orientation determined – at least in part – by genes too?

Research by Thomas J. Bouchard at the University of Minnesota in the 1970s on identical twins reared separately led to some startling revelations that have encouraged the view that genes shape human psychology. These twins, with the same genotype but completely different environments, showed strong concordance in areas such as political orientation, job satisfaction, tendency to divorce, and even in the names of their children, wives and pets.

This led the media, the general public and some scientists to wonder if there were genes 'for' everything. Neuroanatomist Simon LeVay of the Salk Institute in San Diego fuelled the debate when he declared, in 1991, that there were significant differences between the brains of self-declared male homosexuals and men whom the researchers assumed to be heterosexual. LeVay got his data by looking at the brain tissue of the 16 gay men (post mortem) and finding that the neurons in a part of the brain called the hypothalamus were smaller than they were in the 19 heterosexuals.

Many people assumed that these changes were genetic in origin (although they could have equally been caused by other factors – such as unusual exposure to maternal hormones in the womb). Later, other researchers claimed to have found genes linked to male homosexuality.

Meanwhile genes 'for' schizophrenia, manic depressive disorder, intelligence, violence, and alcoholism were apparently discovered. What has not been widely reported is that all these claims have either not been officially published, or have been quietly retracted. This did not stop Daniel D. Koshland Jr, editor of the prestigious journal *Science* arguing that the 'nature/nurture' debate is essentially over and genetic research holds the key to solving society's most intractable problems – such as drug abuse, violent crime and even homelessness.

But British scientist Steven Rose has spoken out strongly against what he calls 'neurogenetic determinism' because it absolves individuals from taking responsibility for their psychological and social identity. LeVay has courted this charge by hinting that people would be more tolerant of homosexuals if it were shown that their sexual orientation was their genetic destiny, rather than a personal choice.

Rose goes on to say the overemphasis on genes lets government and other authorities off the hook too. No need for housing programmes, for example, if homelessness is genetic; then it will be wasted money. Maybe there is a gene for unemployment too, and economic policies are not the reason why people cannot find jobs.

Finally, Rose questions the precious research resources that are being directed towards neurogenetics – when we could use some of the money to find out more about environmental and social factors that are important in people's psychological states. Rose argues strongly against the current US Violence Initiative. This originated in 1992 in a proposal by the then director of the National Institutes for Mental Health, Frederick Goodwin, to identify inner-city children whose alleged biochemical defects may make them prone to violence in later life. Even if a link were found between genes and violence, little would be achieved in the absence of measures to reduce the availability of handguns in the United States.

However, there is still a strong case for teasing out the genes involved in psychological disorders where there is a known genetic component. A major new study has been launched by researchers in Cardiff to find the genes involved in schizophrenia by analysing the DNA of over 200 brothers and sisters who are affected. The team is careful to call these 'susceptibility' genes, rather than genes 'for' the disease.

Schizophrenia affects around one per cent of the population everywhere in the world. An identical twin of a schizophrenic has a 50 per cent chance of developing the disease, which is marked by social withdrawal and loss of drive, delusions, a high suicide risk and – occasionally – violence. It destroys quality of life for most sufferers and their families. In Britain it still accounts for more hospital bed occupancy than any other disease – despite community care policies that have left many patients without adequate care, and have led to several tragic incidents. There have been some intriguing hints as to where the genetic connection may lie. For instance, women with a family history of schizophrenia appear to have an immune system that differs from that of other women. They often have higher levels of a protein called B44 that mounts a strong attack on infections such as influenza. If they get the 'flu when they are pregnant this strong immune response appears to affect the foetus, leading to faulty brain development. The fact that more schizophrenics are born in the spring has been linked to 'flu epidemics. Some schizophrenics show subtle brain damage that is consistent with exposure to maternal infection of this kind. The women show this immune profile because they have inherited a genotype that leads to it. So there is a mother–child influence here: genotype, immune system, infection, brain damage, schizophrenia. A positive step forward would be to identify the women at risk of having an affected child by their immune system profile – looking at their level of B44 – and protecting them from viral infection during pregnancy with immunisation or antiviral drugs. It is an enormous challenge to the Cardiff team to see whether discovery of the susceptibility genes for schizophrenia can truly lead to a better outlook for those suffering from the disease, rather than shifting attention away from the social and environmental factors that are also involved.

Alzheimer's disease (AD) is another important target for the geneticists. This disease of the brain, marked by characteristic deposits of a protein called amyloid, causes loss of memory and deterioration of the personality. It affects five per cent of people over the age of 65, and nearly a third of the over 85s. As people live longer, AD is proving to be a major threat to the wellbeing of older people and their families.

AD is caused by genetic and as yet unknown environmental factors. The latter may include infection, head injury and perhaps aluminium poisoning. Where it affects a person under age 65 (so-called early-onset AD) there is often a family history. Some of these cases have been associated with a mutation in the gene for amyloid, on chromosome 21. It seems that this leads to a production of an abnormal form of amyloid. Amyloid is now known to occur naturally within the body; it is the abnormal form that leads to the disease. There is another mutation that occurs in early-onset AD. This is on chromosome 14 and it is associated with an as yet unidentified gene. But these early-onset cases account for only five per cent of the total. What about the rest? There is evidence that the common form of AD may be linked to inheritance of a certain form of a protein called apoE. ApoE is active in the brain and comes in three versions: E2, E3 and E4. E3 is the commonest, and E4 is associated with the development of AD. People with E4 have a four times greater risk of getting AD than normal. It is important to realise, however, that not everyone with the E4 gene gets AD, and not everyone with AD has the E4 gene. Although there is no cure for AD, genetic research has brought us very much closer to understanding how the disease works.

Mutation and gene hunting

All this information about the genetic basis of various diseases would be useless without tests that can be carried out on individuals' DNA to see whether they carry a particular gene. These tests are becoming quicker, cheaper and more sensitive all the time. Within a few years it might even be possible to have your genes tested at the local surgery using a kit that gives results in minutes.

Much DNA testing is not based on testing for the faulty gene itself, but for linked markers. The linked marker is a bit of DNA – it may not be a gene itself – which is close to the faulty gene on the chromosome and so is likely to be inherited with it. If you have the marker you are likely to have the faulty gene too.

In research, looking for these distinctive bits of DNA in an affected family or population is often the first clue to the gene itself. This is how genes for HD and CF were found. First the markers were located; these then led to the genes. The HD gene became the basis of a test that was made available before the gene itself had been found. When you read claims in the media that the gene 'for' some diseases has been found, more often than not it is a marker that has been found, not the gene itself. Inevitably the discovery of the gene follows within a few months because the marker shows whereabouts on the chromosome to look. So when the popular press announced that a gene 'for' dyslexia had been found by US scientists in late 1994, what the team had actually done was to spotlight a region on chromosome 6 that was distinctive in children with this learning disorder. A gene defect that somehow plays a role in dyslexia may well be found in this region.

What follows is a simplified and basic description of how this knowledge of markers and genes is put into practice in carrying out a DNA test. The aim is to find the marker or gene within the vast expanses that make up the human genome. The first step is to extract DNA from the person being tested. Normally this involves taking a blood sample about 20 millimetres in volume (this is similar to the amount taken for other blood tests). The red cells and the plasma (a fluid containing white blood cells) are separated by spinning the blood sample in a centrifuge. Plasma is lighter than red cells, so it floats on top of the sample after it has been centrifuged.

Next the DNA is extracted from the plasma roughly as described for the onion DNA at the beginning of this book. Usually the extraction will give about 600 micrograms (a microgram is a millionth of a gram). This is enough for several DNA analyses.

Sometimes it is just not possible to obtain a syringeful of this amount of blood – in forensic work, for example, or from foetuses in the early months of pregnancy. However, thanks to the polymerase chain reaction (PCR), it is now possible to start out with a

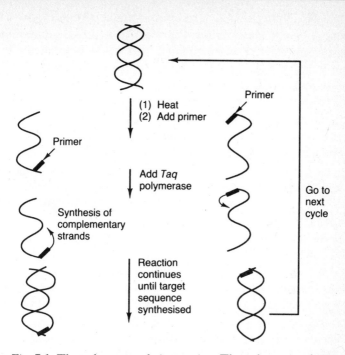

Fig. 7.1. The polymerase chain reaction. The polymerase chain reaction (PCR) is a way of getting a great deal of DNA from a very small sample. Heating the sample splits it into single strands. Each is then tagged with a small complementary sequence known as a primer. This acts as a marker for synthesis of complementary strands using the appropriate mixture of enzyme and free nucleotides. Repetition of the process over and over soon leads to the production of substantial amounts of the DNA sequence marked off by the primer.

much smaller DNA sample and obtain enough for tests by reacting it with a mixture of an enzyme and other chemicals (Fig. 7.1).

The discovery and development of the PCR has been the main advance in DNA technology over the last decade. It came to its inventor, Kary Mullis of the Cetus Corporation in the USA, as a flash of inspiration when driving along a moonlit road in north California in 1983. When, ten years later, Mullis was rewarded with a Nobel Prize for his discovery of PCR, some newspapers explained the technique as a way of 'photocopying' DNA. This analogy is not too far from the truth. But unlike conventional photocopying, PCR is exponential; that is, the number of DNA molecules doubles each time the reaction is carried out, following the sequence 1, 2, 4, 8, 16 . . .

PCR is a clever way of exploiting base pairing. Suppose we have just one nanogram of DNA (a nanogram is a billionth of a gram) that contains a sequence, known as the target, which needs to be analysed. This could be a gene, part of a gene, or some other stretch of DNA such as a marker. If we know the sequence of the target, or at least some of it, two smaller stretches of DNA called primers, which are complementary to the ends of the target, can be synthesised on a laboratory machine (see p. 105).

The DNA sample is heated so that it is denatured – or turned into single strands. After cooling the primers are added. These find and base pair to their complementary sequences, marking either end of the target. This is called annealing. Then DNA polymerase is added, and the two small double-stranded sections, created by annealing, form starting points for the enzyme to replicate each single-stranded target region. The result is two double-stranded molecules of target DNA. This is one cycle of PCR. The rest of the DNA sample remains in the single-stranded form because it is outside the region marked out by the primers.

A second cycle gives four molecules, as each of the two 'daughters' doubles up. A third gives eight . . . and so on. Within a few hours our one nanogram sample of DNA has multiplied about a million times to give a milligram – enough for hundreds of DNA tests.

When Mullis first tried out PCR in the laboratory, however, he uncovered a snag. Every time the DNA was denatured to start off the next cycle, the polymerase was denatured alongside it. Most proteins cannot stand the heat: this is why eggs undergo such dramatic changes of texture when they are scrambled, boiled and fried; the protein molecules in the yolk and white unwind into long chains that tangle with one another to form complicated moist, chewy, or leathery meshes. This is all very well at breakfast, but in an enzyme-based experiment it is a disaster. So in the early days of PCR a fresh slug of enzyme had to be added after every cycle to replace the denatured, and therefore useless, enzyme. This was tedious, time consuming, and expensive. It threatened the development of the technique for widespread clinical use.

Fortunately the problem was soon solved by the introduction of a heat-stable polymerase from a bacterium called *Thermus aquaticus*. This interesting creature thrives in hot springs at temperatures around 100 °C. It can do so only because its enzymes are stable to heat. For humans and, indeed, all other plant and animal species, life is generally extinguished if the temperature of the organism climbs to much over 40 °C, because essential enzymes are denatured.

The new polymerase – *Taq* polymerase or *Taq* for short – proved to be ideally suited to the PCR environment and has earned its discoverers millions of dollars in patent and licensing revenues. PCR is now largely automated: the scientists prepare a mixture of *Taq*, DNA sample, primers and water with some added salts, and place it in a machine that will take it through as many cycles as are necessary to amplify the sample for analysis.

The power of PCR, in the context of human genetics, has been shown through the development of pre-implantation diagnosis. In this technique, eggs and sperm from a couple at risk of conceiving a child with severe genetic disease are fertilised in a small glass dish (this is *in vitro* fertilisation or IVF) and allowed to develop to the eight cell stage. This takes about three days. Then one cell is taken from each embryo and its DNA extracted and multiplied by PCR. This way each embryo can be screened very rapidly for defective genes. Healthy embryos are placed in the uterus and, assuming implantation takes place, the couple can proceed with the pregnancy secure in the knowledge that their baby will be free from the disease. The technique has been used to prevent cystic fibrosis, haemophilia and Tay–Sachs disease, the last of these being a progressive nervous system disorder that is usually fatal by the time a child reaches adolescence.

Whether PCR is needed or not, the DNA analysis that follows is based on a technique called Southern blotting, developed by E. M. Southern. The extracted DNA is first mixed with a cocktail of restriction enzymes, of the type used in genetic engineering, and chopped up into smaller fragments. Next these fragments have to be separated from one another. This is done using a technique called electrophoresis. The DNA sample is applied to a slab of gel. Usually this is made from agarose, a carbon-based polymer obtained from seaweed, which forms a stiff gel when it is cooked with water in a microwave oven. The gel is supported on a tray and then submerged in a buffer, a solution containing various salts that can conduct electricity. An electric current is then applied to the two ends of the gel using a power pack.

DNA has a negative electrical charge, which is carried on its phosphate groups. This means it will move in an electric field. The DNA fragments move along the gel, towards the end that is attached to the positive terminal of the power pack. The smaller the fragments, the faster they can move in the electric field. So at the end of the electrophoresis session the fragments are spread out on the gel according to their size, although at this stage they are invisible.

Then the double-stranded DNA of these fragments is separated into single strands by treatment with an alkali. The reason the process is called blotting is because of the next step – in which the single-stranded fragments are transferred from the gel to a nylon membrane in much the same way as a message in wet ink is transferred to blotting paper. Walk into any molecular biology laboratory and sooner or later you will come across a 'low-tech' arrangement of glass plates, gels, nylon membranes, paper towels and heavy weights, all sitting in a tank of buffer. This is the apparatus in which the DNA fragments slowly soak up into the nylon membrane to which they then adhere, sometimes with the aid of gentle baking in an oven.

The next step is to select a probe that will fish out the gene or sequence of interest from the 'ladder' of DNA fragments on the membrane. Probes, which were discussed in Chapter 3, are strings of nucleotides whose bases are complementary to a section of sought-after sequence. We do not need the whole complementary gene sequence in the probe, just enough to make sure it is unique to the genome. For instance if our gene starts with, or includes, the sequence AAT it is of little use making a trinucleotide probe with the sequence TTA because this will find hundreds of complementary sequences through the genome. For specificity (picking out one sequence only within the genome), a probe at least ten nucleotides long is needed.

The probe has to have some kind of tag that will enable it to be identified later on. Traditionally the probe is built up from nucleotides that contain a radioactive phosphorus atom. This does not affect the chemistry of the probe, but the radiation it gives off will darken photographic paper. Because laboratories have to be specially set up to handle radioactive substances – from monitoring the health of scientists using the probes to working out how to dispose of the waste produced by these processes – much effort has been put into non-radioactive ways of labelling probes in the last few years. Probes can therefore be tagged with fluorescent molecules that will shine brightly when exposed to ultraviolet light, or linked to enzymes that will produce a coloured compound when exposed to a substrate (the name given to the compound with which the enzyme interacts).

The probe is incubated with the single-stranded DNA mixture on the nylon membrane and soon finds its complementary sequence. Note that if the mixture had not been denatured and 'blotted' the probe would not have been able to stick to its DNA target. Until this point, everything on the gel is invisible. The climax occurs when the gel is exposed to a visualising agent. All the bands that bind to the probe shine out. How many bands you see really depends on the kind of test you are doing. If you are looking at a marker you might see just one or two bands. This can be very informative – especially if it is in a family where one pattern is common to non-affected members and the other to affected members.

Suppose, for example, you are trying to build a family tree for a disease where the gene involved is unknown. Somewhere near this gene is a marker of a very common kind, a restriction fragment length polymorphism or RFLP. This is a natural variant in DNA sequence (a polymorphism) that either creates or abolishes a site where a restriction enzyme can act. For example if GAATCC occurs in some people at the same point in the genomes as GCATCC occurs in others, this is a polymorphism. On its own it is quite harmless because it probably occurs outside a gene (remember the majority of our genome is 'junk' DNA). However, the restriction enzyme *Eco*RI snips DNA between the G and the A of the first sequence, but cannot cut the second sequence at all. So if you incubate the DNA sample with *Eco*RI the band containing the probe corresponds to a shorter DNA fragment if the first polymorphism is present than if the second one is. It will therefore appear on a different position in the gel. If this first polymorphism is inherited with the disease, the band corresponding to shorter DNA is indicative of inheriting the faulty gene, whereas the person with the band corresponding to longer DNA is in the clear. That is the principle; obviously the strength of the correlation depends on how close the marker polymorphism and the gene involved actually are. It takes research with DNA from hundreds of people to confirm these connections. If the actual gene involved in a disease is known – for example factor VIII in haemophilia – then testing will focus on pulling out the gene itself with a probe and perhaps searching for actual mutations within it.

DNA testing is the fastest-growing area in medical diagnostics. According to the US Office of Technology Assessment, the number of such tests will increase tenfold over the next decade. The aims and achievements of DNA testing vary. Programmes to establish carrier status for relatively common disorders such as CF and Tay–Sachs disease are now widespread. In Sardinia, such a programme has already cut the rate of thalassaemia from one in 250 to one in 1200 over the last 20 years. If couples who are both carriers decide to have children, they can choose pre-implantation diagnosis, as described above. Or they could opt for prenatal diagnosis during pregnancy – followed by abortion if the foetus is found to be affected. A third choice would be to go ahead with the pregnancy, with or without the knowledge of the baby's genotype. Tests for Down syndrome, based on chromosome analysis rather than DNA, have been available for many years, and the newer DNA tests raise the same mixture of ethical and economic issues. The people who advise the couple – doctors, nurses and counsellors – are supposed to guide, but not influence them. Yet in the long run, testing programmes will only be cost-effective if most affected foetuses are aborted rather than allowed to survive as an expensive burden on health care resources. How are medical personnel to stop such considerations influencing the advice and information they give to their patients?

Some people have suggested that widespread prenatal (including pre-implantation) diagnosis signals a trend towards eugenics, the improvement of a population by the application of genetics. Another concern is that the possibility of elimination of genetic disability might lead to the worsening of the position of disabled people in society. In addition, prenatal diagnosis has a positive moral aspect in that it attempts to reduce the potential suffering of the foetus (if it is not born, it cannot suffer) and its family. All these issues need to be thoroughly aired as carrier and prenatal testing becomes more widely available. However, there is one scientific point that can be clarified immediately. We will *never* eliminate genetic disease. By now you will have an appreciation of the fluid and dynamic nature of the genome. New mutations happen all the time. In the last few years, databases of mutations found in people suffering from single gene defects have been compiled in an attempt to understand the molecular nature of the disease. What emerges from these is that a significant proportion of the single gene defects are new mutations – there is no family history of the disease. Instead, the patient has just become the 'founder' of yet another family suffering from it. Somehow the patient's germ cells have been mutated during their lifetime, and the change will be passed on.

As more and more genes roll off the human genome production line, and researchers come to grips with the complexities of common multigenic disorders, the prospect of screening large numbers of the population will become more than a distant reality. For many of us, the idea of having our whole genome 'taped' is rather a frightening possibility. Who will have access to the information, and what use will they make of it?

We share our genes with our family, so genetic data nearly always have implications for someone else. In a recent survey in the USA, 60 per cent of those questioned thought such information should be available to (in decreasing order of importance) partners, other family members, insurers, and employers. From a practical point of view it will be many years before the prospect of a 'genetic passport' becomes a reality. There are good scientific reasons, at present, for questioning the usefulness of such a document. For example, if someone is shown to have a gene for susceptibility to heart disease (like the defective ACE gene mentioned earlier), there is really no way of knowing how effective adopting a 'healthy' lifestyle would be for that person. On the other hand, if someone did not have that defective gene, would it be a licence for them to drink, smoke and eat fatty food? Many scientists argue that until we have at least some of the answers to how genes and the environment interact it is unethical to offer DNA testing (or screening, if the tests are applied to a large population) for at least the multigene disorders such as heart disease.

We may soon have the chance to put some of these questions about screening to the test. Breast cancer is a major killer of women in developed countries. Around 28000 women get breast cancer every year in the UK, and 180000 in the USA. Five to ten per cent of these come from families with an obvious history of the disease. For instance, a woman whose mother and sister had or have breast cancer – especially if the disease appeared before they were 50 – will fall into this category. In 1990 Mary-Claire King of the University of California predicted that a gene for familial breast cancer, named BRCA1, would be found in a particular region on the long arm of chromosome 17, close to a marker that was inherited in the disease in several families.

Now BRCA1 has been found, not by King and her team, but by Mark Skolnick and co-workers at the University of Utah. The gene is responsible for perhaps half of the inherited breast cancers. Some families are plagued by breast and ovarian cancers, and defects in BRCA1 probably account for most of these. A woman with a mutated BRCA1 has a lifetime risk of about 90 per cent of getting breast cancer. For the general population the risk is just under 10 per cent.

It looks as if there are several genes involved in familial breast cancer. British and US teams have located a second, BRCA2, on the long arm of chromosome 13 (the actual map location is q12-13 – for a translation, see p. 61). They do not yet have the gene itself, and there are about 100 genes in all in the target region. Mutations in this gene probably account for most of the other inherited breast cancers.

At least one of the breast cancer genes having been discovered means that more accurate tests can be developed, based on the gene itself and its mutations rather than nearby markers. Already, though, questions are being asked about how useful such tests will be. Women with a family history of breast cancer already know they are at risk. What they may not appreciate is the magnitude of the risk, which depends on the number of family members affected and how old they were when they got the disease. Risk varies with age too – the older the woman, the more likely she is to escape the disease. So advice and counselling have to be tailored very much to the woman and her family.

What more can a DNA test add? Discovery of the specific mutation involved in each case (there are several different ones that have been found already) may give far more precise estimates of the risk of cancer developing. Once the risk is established there is the option of regular mammography for early diagnosis, enrolment in a preventative programme of tamoxifen treatment (as discussed in Chapter 2) or even prophylactic removal of a breast. No-one can yet say how effective such measures might be.

You might think that having found the genes involved in familial breast cancer must mean an immediate leap in understanding the other 90 to 95 per cent of breast cancers. Surprisingly the majority of women with breast cancer who do not inherit it do not have mutations in BRCA1. This is disappointing, to say the least, and has led some researchers to wonder whether BRCA1 is really the right gene. It certainly makes the case for screening rather shaky.

Even so, Skolnick has set up a company, Myriad Genetics, to commercialise his research. This means developing a test for BRCA1. To fund this he has applied for a patent on the gene. Inevitably there has been some opposition. Some scientists fear it will lead to secrecy and delay in sharing important results. Others just think it is wrong to patent parts of the human body, and previous attempts of this kind have failed. One thing is sure – the fate of Skolnick's application will have a huge impact on the prospects of patenting other human genes.

DNA and identity

Your identity is written into your genome, and this fact is the basis of a powerful group of technologies popularly known as DNA

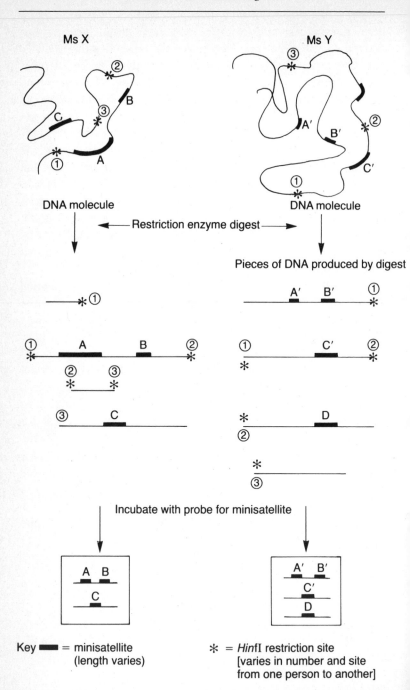

Key ■ = minisatellite
(length varies)

∗ = *Hin*fI restriction site
[varies in number and site
from one person to another]

fingerprinting. Professor Alec Jeffreys of Leicester University was the first to apply Southern blotting analysis to establish identity on the basis of an individual's DNA.

As we saw in Chapter 3, human DNA has a fairly distinctive molecular landscape, with its repeats, exons, introns, transposons and other bits and pieces. Among all this Jeffreys picked out features called minisatellites – stretches of DNA where the same bits of sequence, usually up to 20 bases, would repeat themselves. What excited Jeffreys was how much the lengths of the minisatellites varied from one person to another. One person might have minisatellites hundreds of times longer than her friend next door, for instance. Given this variability, and that fact that the minisatellites are scattered throughout the genome, it make sense to try to visualise minisatellite patterns from different people's DNA and use them as a kind of fingerprint (Fig. 7.2).

Fig. 7.2. DNA profiling. Two women, Ms X and Ms Y, have volunteered for DNA profiling. First their DNA is extracted from a sample of blood or saliva. This DNA will be studded with so-called minisatellite DNA – bits of repeated base sequence that will vary markedly between the two women. Here the minisatellite DNA is indicated by letters A, B etc. To identify the differences between the minisatellite DNA of Ms X and Ms Y, the samples are first of all chopped up into smaller bits using an enzyme called *Hin*fI. This part of the process is called a restriction digest. The bits of DNA, labelled 1, 2 etc., are from the enzyme chopping up the samples at the points marked with asterisks. Some of these bits contain minisatellite regions. The bits are separated from one another by size on a slab of gel using an electric current that makes the small bits move along faster than the larger bits. This part of the technique is called electrophoresis.

After electrophoresis, the slab of gel is incubated with a solution that contains a piece of DNA known as a probe. The probe is complementary in its base sequence to the minisatellites and, as it is tagged with a marker such as a radioactive atom or fluorescent molecule, will pick them up on the gel. So the size and number of minisatellites in Ms X's and Ms Y's DNA can be revealed by this technique.

First the DNA is chopped up, in the usual way, using the restriction enzyme *Hin*fI (so called because it comes from the bacterium *Haemophilus influenzae*). This recognises and clips DNA wherever the sequence GANTC appears in the genome (the nucleotides have their usual abbreviations and N stands for any nucleotide). *Hin*fI is the standard enzyme used for fingerprinting in Europe. It helps to use the same enzyme in each laboratory, so that results can be compared more easily.

Some of the fragments produced by this treatment contain one or more minisatellites, and these can be visualised, after blotting, with a probe that finds a piece of sequence within the minisatellite. But these species will vary in length, from person to person, because their minisatellite lengths vary. What appears after probing is a set of up to 20 bands that looks rather like a supermarket bar code. Our two friends might have 'fingerprints' that look as different as, say, the code on a packet of cheese and the code on a tube of toothpaste at the supermarket checkout.

This technique has proved very useful in sorting out paternity and immigration disputes. It can show, for example, whether a man wanting to bring his wife and children into the country is really the children's father or whether he is trying to smuggle in his brother's family. All you have to do is run DNA from each family member side by side in the same Southern blotting experiment. Everyone inherits half their bands in this test from their mother, and half from their father. It is usually easy to tell, by eye, who is and who is not a parent of a particular child.

Jeffreys points out that since its first use in an immigration dispute in 1985, DNA testing has shown the claims of the majority of immigrants to be correct. An unfortunate spin-off from the technique, however, is that it will also disclose non-paternity, which usually means that the father of the child is not the husband of the mother. According to scientists involved in DNA testing this is commoner than they might have expected – and of course it requires extra tact when talking through the results with the family that has just taken the tests.

DNA fingerprinting (as it is still called, although the term DNA profiling is now preferred by those in the business) has obvious applications to forensic work. The DNA fingerprint of an individual is the same in all body tissues. So you would expect blood from a guilty suspect to give the same pattern as, say, his saliva recovered from the scene of a crime. But the patterns are complex and hard to compare when produced in different laboratories at different times (inevitable unless the suspect is arrested immediately after the crime). It is also hard to log the fingerprints into a computer database. Finally, the technique is not very sensitive, so it is not really suitable for either tiny or mixed samples (such as semen mixed with vaginal secretions in a rape case).

So for forensic work another method has been developed, in which a probe for just one minisatellite is used. This gives two bands per sample – one for each version of this minisatellite inherited from each parent. Because of minisatellite variability we need test with only four different probes to get a near unique pattern for each individual.

The world's first DNA profiling success in the forensic area was the solving of a double rape and murder in Leicestershire in 1986. The case was particularly dramatic for two reasons. First, a man who had confessed to the second murder was cleared because his DNA did not match that from the scene of the crime (DNA profiling should always eliminate the innocent, unlike some other forms of forensic testing). Second, the guilty man, Colin Pitchfork, persuaded a colleague to stand in for him when DNA testing of all the men in the area was organised. It was only when the police were tipped off about the deception that Pitchfork was arrested, tried, and given a life sentence.

DNA profiling is used only for serious crime – rape and murder – and in England and Wales the potential for its use is around 2500 cases a year (out of 5.5 million offences recorded by the police in 1992, the last year for which figures are available). Inevitably PCR is being used in forensic work, and extends the range of samples that can be analysed to, for example, the saliva on the back of a postage stamp.

However, DNA profiling has had a rough ride in some courts, and in some cases has been rejected altogether as a source of evidence. The reasons are both technical and statistical. It is all too easy to get poor quality results with any kind of DNA analysis. You are working with enzymes and with DNA, both of which are sensitive to changes in temperature and chemical environment. Techniques such as Southern blotting that involve many stages are subject to human error. Even if you were to carry out the technique under the most carefully controlled conditions, the chances are that there would be a small amount of variability of the positions of the bands in the final pattern produced by the test. This would happen whether different people did the test on the same day, or you did it yourself on two different days – or even in two different laboratories.

But scientists can get together and set standards that take account of the possibility of all these sources of inaccuracy. In the early days of DNA fingerprinting, this was not always done. Forensic experts say these technical errors favour the guilty by not showing up matches that should occur.

Perhaps more difficult to address is the common belief that a DNA profile is unique. Surveys done on data from samples of the population suggest that this is almost certainly true; yet it remains a point of contention, with criticisms that the near matches of people from close communities are not taken into consideration. Some scientists have suggested that if eight, rather than four, probes were used in the analysis the likelihood of 'false positives' would be eliminated.

Whatever the drawbacks of DNA profiling, the British authorities have decided to use it to create a database of people convicted of imprisonable offences. This idea is also being taken up in the USA, where individual states have been running such a programme for some time, and have achieved arrests and convictions for several crimes using the data that it provides. The data would be stored digitally on an optical disc, and it is not clear whether data from people subsequently proved innocent could be erased. The plan is that it would be scanned with data obtained from the scenes of major crimes.

Civil liberties groups say the database should be confined to sex offenders and murderers and ask that people on the database should be able to check it for themselves. They also argue, as do most people in the forensic field, that no-one should be convicted on DNA evidence alone.

8

New genes for old

Increasingly, medical research is dominated by DNA. The hunt is on for genes that can help to answer the really big questions, such as how does a single cell grow and develop into a complex body, and what *really* causes cells to grow into tumours sometimes. At the same time new DNA-based drugs and therapies are making their way from the research laboratory into the hospitals and general practitioners' surgeries.

Life, death and the cell

In California, there is a group of people who believe humans are meant to be immortal and that our bodies can last indefinitely if we only have the right psychological outlook! There is absolutely no scientific evidence for this, but the belief that the cells that make up our bodies are immortal was certainly once very widely accepted. Now our view of the life expectancy of cells has changed completely. According to the latest (and still controversial) research, the natural state of our cells is death, and only the constant prodding of genetic signals actually keeps them alive.

The myth of cell immortality originated with Nobel Prize winner Alexis Carrel, a French surgeon who was interested in organ transplantation and tissue culture. In 1912, he started to culture some cells from a chick's heart to see how long they would survive outside the animal's body. When they outgrew their culture vessel, they were divided up and transferred to new vessels – a process known as subculturing. By 1922, they had been subcultured 1860 times and the press and public began to take an interest. According to one reporter writing in New York, there were enough cells to form 'a rooster . . . big enough to cross the Atlantic in a stride . . . so monstrous that when perched on this mundane sphere, the world, it would look like a weathercock'. The cells outlived Carrel, who died in 1944. In the end they did not die, but were discarded in 1946. No wonder the scientists thought cells were immortal. It was being in a body that made cells die, they argued.

Then in 1961, Leonard Hayflick and P. Moorhead showed that Carrel was wrong. They grew a number of different kinds of human cell in culture and showed that they always died after about 50 cycles of subculturing. The older the cells were, the fewer cycles they went through before they expired. At first the scientific establishment refused to believe Hayflick and Moorhead, hinting that some contaminant must be killing off the cells. However, the pair persisted and when their work was finally published it became a landmark in the understanding of human cell biology. As for Carrel's famous culture, it now seems that cells did indeed die from time to time, but so convinced were Carrel's team that the cells were immortal, that they assumed the deaths were due to accidental contamination – so they merely 'topped up' the culture with fresh cells.

We now know that cells do sometimes become immortal, dividing endlessly to form a tumour. These cancer cells are said to be transformed, but normal cells have a finite lifetime. Some, like brain cells, never divide at all. Others, like the cells that line the digestive system, are renewed twice a day. Once our bodies have reached maturity, with about ten thousand billion (10^{13}) cells, then cell division reaches a sort of equilibrium where new cells are produced to replace those that wear out or become damaged.

It may sound from this as if the body could carry on indefinitely after all, simply producing new cells when necessary. However the continual biochemical activity inside the cells has its price tag. Although people worry that synthetic chemicals such as food additives and pesticides could give them cancer, the worst inner pollutant is a completely natural substance – oxygen. Although we cannot survive without oxygen, it is at the same time extremely toxic. Not in itself, perhaps, but because it forms a number of related substances known as reactive oxygen species (ROS), which attack vital molecules such as DNA and proteins in our cells.

American cancer scientist Bruce Ames reckons that ROS make around 10000 hits a day in each cell of the human body. Fortunately, most of this damage gets repaired by enzymes, in a fashion similar to mutation repair, as discussed in Chapter 7. However, the enzymes become less efficient as time goes by and inevitably this so-called oxidative damage accumulates, leading to tissues and organs that do not function correctly. The result is the so-called degenerative disorders such as Alzheimer's disease and Parkinson's disease.

Tom Kirkwood, the UK's first Professor of Biological Gerontology (the scientific study of ageing) has developed the 'disposable soma' theory to account for why we age. The maximum human lifespan is around 120 years. This probably has not changed much over time, although decreasing rates of child mortality in developed countries have led to a dramatic rise in average life expectancy over the last few hundred years. The reason we cannot expect to survive beyond 120, says Kirkwood, is that cellular repair and maintenance comes expensive in terms of biochemical energy and each of us has only a limited biochemical budget at our disposal. We have to divide our budget between reproduction – essential for the survival of the species – and keeping our body (soma) going. The resources available for the latter are just not sufficient to allow indefinite survival.

Genes play a part in longevity, but no-one yet knows what these genes are, in a human. Long-lived mutants have been found in fruit flies and nematodes (worms). For instance, flies with unusually active genes for the enzyme superoxide dismutase (SOD) have prolonged lifespans. Cynthia Kenyon and her team, working in California, recently reported on a worm with a mutation in one particular gene that gave it a lifespan equivalent to a 200-year-old human. But the concept of manipulation of the human version of these genes remains in the realm of fantasy. A more practical way of having a long and healthy life might be to boost the cell's defences against ROS, simply by eating more fresh fruit and vegetables. These contain vitamins that can intercept the ROS before they can do any damage to the cell. Another solution might be to eat less! Rats on a calorie-restricted diet live longer than those on a normal diet. It makes sense because the less food there is to be burned to make energy, the less oxygen is used, and the fewer ROS there will be in the cell.

There is also increasing emphasis on the state of mitochondrial DNA as a marker of ageing. Mitochondria are the sites in the cell where most energy production takes place. They do not seem to have the same protective mechanisms for repairing their DNA as the nucleus does – instead new mitochondria are made to replace damaged ones. As damage accumulates, energy production in the mitochondria becomes less and less efficient. Once the power-house of the cell starts to run down, death is inevitable.

The quality of later life could also be improved by extending the number of times a cell divides. Animals with short lives, such as rats and mice, which live for only a year or two, have cells that divide around ten times. But the cells of a Galapagos tortoise, which has a lifespan of up to 200 years, divide about 100 times. So there appears to be a connection between lifespan and number of cell divisions. Every time a cell divides, the ends of the chromo-somes, which are known as telomeres, get shorter. Scientists in California have found drugs that protect the telomeres from this shortening and seem to rejuvenate the cells by prolonging the number of cell divisions.

Genes and the unfolding body plan

At the other end of life, genes are involved in shaping the body from the formless mass of cells that is the early embryo. Most of what we know about genes involved in development comes from experiments on fruit flies (*Drosophila*) – especially those carried out by Edward B. Lewis and his team at the Californian Institute of Technology over the last 40 years. Occasionally fruit fly mutants occur in which body parts are displaced – a leg in place of a wing, for instance. The mutations were found to occur in genes in the so-called HOM group. HOM genes, in ways which are not fully understood, direct cells in the developing embryo to the right place. A HOM gene 'tells' a cell it is meant to be in the fly's thorax, leg, or abdomen. Different HOM genes take charge of different parts of the body.

Advances in DNA technology have contributed a great deal to the study of development. For instance, it has been possible to find out which cells in the fly embryos contain active HOM genes by looking for RNA in the cells. This is a sign that the gene is switched on and working. If the gene is inactive, it will not produce any mRNA. There is a full set of HOM genes in each fly cell, of course, but it appears that different HOM genes are activated in successive sets of cells along the long head-to-tail axis of the body. If one HOM gene is missing or defective, it appears that one of the others can take over – leading to the development of a different body part in that position.

All invertebrates have HOM genes. The corresponding genes in vertebrates are called *Hox* genes (see p. 130). HOM and *Hox* genes have been analysed from a number of species, including mice and humans and it turns out that they all have remarkable similarities in their sequences. They are all marked by a feature called a 'homeobox' – a part of the sequence that codes for a protein region called a homeodomain. This is known to bind to control sequences in DNA. So it looks as if HOM and *Hox* genes make proteins that can turn on other genes – exactly what you might expect from genes involved in development.

According to William McGinnis of Yale University, a human *Hox* gene can function even in a fly. The gene is like the one that causes head abnormalities in flies if it is switched on in the wrong cell. McGinnis made the human gene switch on in all fly cells, and found that similar abnormalities resulted. When the fly received the corresponding mouse gene, it developed a structure more like a thorax where the head should have been, with front legs in place of antennae.

It is a humbling thought, but it appears that flies, worms, mice and humans are all far more similar at the DNA level than you would ever suspect from their different outer appearance. The same kinds of genes and molecular mechanisms appear to control the unfolding architecture of the three-dimensional mature organism. The genes derive from a common ancestor – an ancient worm-like creature – and have remained similar over at least 700 million years. And there are corresponding genes in plants that control the development of their form.

Understanding how these genes work should shed some light on how birth defects occur and may form the basis of new kinds of treatment for people affected by them.

Genes and cancer

It is now clear that cancer, which affects one person in three in developing countries, is a disease of damaged DNA. Unlike the inherited diseases discussed in Chapter 7 most cancers arise from somatic, rather than germline, mutations that are acquired during the lifetime of an individual.

The hallmark of cancer is uncontrolled cell division, resulting in a tumour or 'growth'. So understanding the molecular basis of cancer means understanding the intricacies of cell division and the genes that control it.

The first clues to the genetic basis of cancer came from the discovery of viruses that can cause tumours when they infect animal cells. The viruses carry genes known as oncogenes, which seem to have the ability to turn normal animal cells into cancer cells. Michael Bishop and Harold Varmus then showed that the viral oncogenes (v-oncogenes) have counterparts in animal cells; these they called cellular oncogenes (c-oncogenes) or proto-oncogenes.

Put simply, v-oncogenes are corrupted versions of c-oncogenes. Viruses pick up oncogenes from their hosts during infections. There are many different oncogenes scattered around animal genomes. They code for all kinds of different proteins that are used in the cell's growth and division cycles. If a virus introduces a defective oncogene, then the delicate molecular machinery involved in keeping a check on cell division may be upset. An abnormal protein may be produced, for instance, that could lock the cells into an eternal loop of division.

In fact viruses do not seem to be a major cause of human cancer, although they are important in the development of cervical, penile and some liver cancers. The discovery of oncogenes, however, was certainly a major signpost to important new insights. For instance, point mutations in the oncogene *ras* have been found in some bladder, lung and colon cancers. Oncogenes can also be amplified: in neuroblastoma, a tumour of the nervous system, the presence of many copies of the oncogene *N-myc* is associated with a poor prognosis.

Sometimes c-oncogenes malfunction when they are transported out of their normal genomic environment by chromosomal translocation. The Philadelphia chromosome, which was discussed in Chapter 3, involves chromosomes 9 and 22. The oncogene c-*abl* moves from 9 to 22 during the process. Once on chromosome 22, it is transcribed and translated into an abnormal protein that is associated with the development of a form of leukaemia. These and other rearrangements are easily detectable by looking at the chromosomes under a microscope using new staining techniques. For instance, 'chromosome painting' uses a kit of coloured dyes that allow the visualisation of the region that contains c-*abl* in its new and dangerous environment; this is proving to be a very precise tool for both diagnosis and prognosis.

In the last few years another group of genes involved in cancer has been uncovered. These are called tumour suppressor genes and code for proteins that can put the brakes on cell division. These genes act in a dominant fashion; that is, if one allele in the pair is mutated or inactivated for some reason, the other can still protect the organism. But if this remaining allele becomes inactivated by mutation the result is complete loss of the tumour suppression facility. This inactivation of a second allele is called loss of heterozygosity. The dominant way in which the tumour suppressors act has led to the development of the 'two-hit' hypothesis, which suggests that two mutations are needed – one in each allele – before there is a danger of cancer.

The two-hit hypothesis explains why some cancers have a family history. The argument goes that the first mutation is inherited, a second mutation acquired during life will open up the possibility of a cancer developing. This is the case in the inherited breast cancers discussed in the last chapter. A particularly important tumour suppressor gene called p53, which was discovered by David Lane of the University of Dundee, lies on chromosome 17. Lane calls it 'the guardian of the genome' because it swings into action when cells are damaged, delaying cell division long enough for the damage to be repaired. The fact that most human cancers seem to be associated with mutations in p53 underlines its importance. The gene was even voted 'Molecule of the Year' by the journal *Science* in 1993.

Perhaps the hottest new topic in cancer genetics is programmed cell suicide, or apoptosis. According to Martin Raff of London University, it is natural for cells to die off and only signals from neighbouring cells keep them alive. What could be happening when cancer develops is inhibition of apoptosis, perhaps by the 'reprieve' signal being turned on when it should not be. In other words, cancer happens when cells fail to die – rather than when they grow and divide out of control. Backing for this idea comes from the discovery that *bcl*-2 is indeed damaged in some forms of cancer and may therefore be expressing itself inappropriately.

There are bound to be many more genes involved in cancer, for example those that code for enzymes repairing DNA damage. So where does this leave prevention and cure? The DNA damage that leads to cancer is similar to that occurring in ageing, so the same precautionary measures can apply. High intake of vitamins through fresh fruit and vegetables has already been shown, by cancer researchers, to cut the risk of cancer – and further huge studies of the link between cancer and diet are continuing.

Despite popular belief, pesticide residues and other synthetic chemicals that are widespread in the environment are not major causes of cancer – although anyone exposed to industrial levels of these may be at risk, and should stick closely to health and safety guidelines for handling these substances. As Bruce Ames has pointed out, natural substances such as mushrooms and celery contain far more potent carcinogens than most industrial chemicals. Most researchers agree that the best way to damage your DNA and get cancer is to smoke. And the best prevention is to stop!

Gene therapy and drugs from DNA

Gene therapy involves giving someone a 'normal' copy of a 'defective' gene to take over its function. Some scientists are already saying that gene therapy will cure AIDS, cancer and heart disease. For the moment it is being tested for a number of single gene defects and in a few cancers.

Gene therapy was first tried in 1980 by an American scientist called Martin Cline in a desperate attempt to save two patients suffering from terminal thalassaemia. The attempt to insert normal globin genes did not work – the patients died and Cline was in a great deal of trouble for carrying out unauthorised experiments.

The road to the first official attempt at gene therapy was a long one. In 1990 W. French Anderson and his team at the US National Institutes of Health treated, with genes, a four year old girl suffering from a rare disorder called adenosine deaminase (ADA) deficiency. ADA deficiency leads to severe immune system problems. Genes coding for ADA were inserted into blood cells taken from the girl (using genetic engineering techniques similar to those described in Chapter 5). The modified cells were then transfused back into the girl's body. Soon her immune system was restored to its normal state, as the new genes began to turn out copies of the vital ADA enzyme. Now several other children around the world have benefited from the same treatment. Meanwhile gene therapies for cystic fibrosis (CF) and some cancers are being tried in other patients, and many more are at the laboratory stage.

Gene therapy is quite unlike chemotherapy (treatment with drugs) in the way it develops from the research laboratory to clinical trials in patients. When a new drug, such as an antibiotic or sleeping pill, is invented it goes from synthesis in the laboratory (or fermenter) to test tube experiments and animal trials through to scale-up in a factory and large-scale human testing. In gene therapy, the gene of interest has first to be identified and cloned. A vector suitable for inserting the gene into cells is then selected. Next comes the creation of a transgenic animal, suffering from the genetic disease to be treated (as discussed in Chapter 6) so the gene therapy can be tested for effectiveness. Finally comes the move from animals into humans.

Already there are many genes that could be used in gene therapies – from those that lead to single gene disorders such as ADA deficiency and sickle cell anaemia to tumour suppressor and developmental genes. The major challenge is how to marry up the new genes with their target cells. The answer lies in selecting the right delivery vehicle or vector. At first, viruses seemed to be the most obvious vectors, because they naturally infect cells, inserting their own DNA into it. Viruses have long been used as vectors in genetic engineering (see Chapter 5) so the technology was fairly well established.

French Anderson used a retrovirus to deliver the ADA genes. Retroviruses – of which HIV is one example – contain RNA, which, as discussed in Chapter 2, is transformed into DNA, which then integrates into the host nucleus. If the retrovirus is carrying a therapeutic gene alongside its own genes, this enters the nucleus of the cell too. One problem with retroviruses is that they infect only rapidly dividing cells, which makes them fine for delivery of genes to blood, liver, and cancer cells but useless for other targets.

CF was always going to be one of the first diseases to be treated by gene therapy because of the intense effort that has gone into discovery of the molecular defect that causes it, as well as its relatively high frequency in the population, but here the gene needs to be delivered to the cells lining the lungs. These cannot be infected by retroviruses, and they cannot be removed from the body for treatment in the laboratory. US scientists chose a different virus, adenovirus (one of the common cold viruses), because it naturally affects the cells lining the nose, throat and lungs. So the CF patients inhale, as a spray, adenovirus carrying a normal CF gene. The trials show this approach is effective, but with the drawback that adenovirus does not integrate into the host's genome, so treatment will have to be repeated from time to time as the new genes are lost from the cell.

A third virus that is making a name for itself in gene therapy is the herpes simplex virus. Known also for its ability to cause painful cold sores, this virus naturally infects cells in the nervous system.

This opens up the possibility of treating some quite common neurological disorders. For instance, in Parkinson's disease there is a deficiency in a brain chemical called dopamine. If you could insert the gene for the enzyme that produces dopamine into the herpes simplex virus, then it might be possible to boost the brain's dopamine production. A cell-based strategy for Parkinson's disease has already been tried – transplanting dopamine-producing cells from aborted human foetuses into patients' brains – with mixed results. Transplanting the gene alone, rather than cells with all their contents, is less fraught with ethical problems, but these ideas are still at the experimental stage.

However, even though viruses can be very efficient vectors for gene therapy – for instance, retroviruses will typically infect all the cells exposed to them with the new gene – they are not ideal in every case. Although the genes that make the virus replicate are removed prior to inserting the new genes, there is always a remote possibility that it will mutate, or pick up some extra DNA in the body and change back into a form that can reproduce itself. If you consider how cold, flu and HIV viruses change over time, you will understand why many scientists are very keen to develop non-viral vectors that behave more predictably.

A British team, based at the Royal Brompton Hospital, London, and St Mary's Hospital, Edinburgh, is tackling CF using liposomes, rather than viruses, as vectors. Liposomes are fatty globules that fuse with cell membranes. So if the normal CF genes are wrapped up in liposomes they stand a good chance of being delivered direct to the cell. What the scientists have done is to develop a spray of liposomes that CF sufferers will be able to inhale to get the gene they need into their lungs. The Edinburgh team have already created a CF mouse on which they tested the technique. Now the London-based scientists have built on these results and have launched a Phase 1 clinical trial. This involves testing the liposome spray on the lining of the nose, where the cells are very similar to those lining the lungs. The results should give the answer to two key questions: is the treatment safe and does it work? So volunteers will report side-effects, and cells from their nose lining will be analysed to see whether they are expressing the normal protein that is lacking in CF patients. The scientists are quick to pay tribute to these volunteers – not only do they have to make repeated hospital visits, they also have to put up with samples being taken from the nose lining, as well as any unexpected side-effects. Without their efforts, however, no progress could be made to help the thousands of young people whose lungs are being destroyed by the disease.

There are always risks to humans involved in clinical trials. One case involving gene therapy recently caused quite a stir in the medical press. A 28 year old French Canadian woman was desperately ill with familial hypercholesterolaemia (FH), a genetic disorder where there is a defect in the gene that makes a protein called the low density lipoprotein (LDL) receptor. This protein is found on the surface of cells, and its job is to drag cholesterol circulating in the blood inside the cells. High levels of cholesterol in the blood lead to fatty deposits, blockage of arteries and heart attacks. The lady in question had a heart attack when she was 16, major heart surgery at 26 and was unlikely to survive many more years. Although retroviruses carrying lifesaving LDL genes were prepared, they could not infect the liver directly. So cells had to be taken from the woman's liver, treated, and then returned to her body – a very risky business, considering she was already quite sick. No-one is sure how well the procedure has worked. So far, her blood has shown a modest reduction in circulating cholesterol, and surprisingly, she has tolerated the surgery well. Delivery of new genes to the liver would help in many other inherited disorders, so either surgical techniques will have to be refined, or vectors that home in on the liver will have to be developed.

The horizons of gene therapy are being extended all the time. Infectious diseases, such as AIDS, might seem to have little in common with single gene defects. Yet genes can be used to attack HIV and stop infection from spreading. One way is to use antisense DNA, as discussed in Chapter 2, to bind to mRNA so that genes that help the virus to infect cells cannot be expressed. In addition, approval has just been granted in the USA for a trial of a novel way of boosting the immune system. The idea is to insert a gene for the protein that appears on the outer surface of HIV – the so-called 'envelope' protein – into a patient's cells. When these are returned to the body, a strong immune response should be produced to the envelope protein (which is, in itself, harmless because it is only a tiny fragment of the complete virus). This is a clever variation on conventional approaches to producing a vaccine against HIV, many of which are based upon the envelope protein because it is known to stimulate the production of antibodies.

There are several ways in which gene therapy could be used to treat cancer. One of the most promising is virally directed enzyme prodrug therapy (VDEPT) which is a way of targeting anti-cancer drugs specifically to cancer cells. One of the main problems with conventional cancer chemotherapy is that the drugs used harm normal cells, as well as the cancer cells. This leads to side-effects – nausea, hair loss, exhaustion – that makes some cancer patients feel that the cure is as bad as the disease. With VDEPT a viral vector is constructed that contains the gene for an enzyme acting on a substance called a prodrug, and converts it into an active anti-cancer drug. The prodrug has no activity until the enzyme acts on it. The enzyme gene is attached to a switch or promoter (see Chapter 2), which ensures that expression occurs only in cancer cells. So when the viral vector arrives at the cancer cells, the enzyme gene is switched on, and it produces the anti-cancer drug from the prodrug (which has been administered in the usual way). Meanwhile normal cells are unaffected because the enzyme can work only in the cancer cells.

Already VDEPT is being developed to dose liver, breast and cancer cells with potent cell-killing drugs. At Hammersmith Hospital in London, for instance, Karol Sikora is using the technology to convert an antifungal drug called 5-FC into a closely related anti-cancer drug, 5-FU. The gene for the enzyme that makes 5-FU from 5-FC is attached to a promoter that switches the gene on only in cancer cells.

Another way of attacking cancer with genes relies on the new dis-coveries about the molecular bases of the disease which were dis-cussed above. First, antisense DNA is being developed to damp down the expression of the oncogenes that are active in many cancers. Second, if tumour cells are given a copy of the p53 tumour suppressor gene, their rate of multiplication plummets. Finally it might be possible to make cancer cells kill themselves off by apoptosis, if the gene *bcl*-2, the so-called reprieve signal discussed above, could be blocked for long enough, perhaps also by antisense technology.

It is important to realise that all gene therapy to date is on somatic cells, rather than germ cells. So the individual alone is being treated and any changes will not be passed on to future generations. Human germline gene therapy, where the genes of all egg and sperm cells would be corrected, is technically almost impossible. It has also been outlawed by all the authorities who control gene therapy and other new medical treatments. But we have in any case already got the technology to remove defective genes from a family – by pre-implantation diagnosis, as discussed in Chapter 7.

Currently, in Britain, gene therapy trials need extra permission from a government appointed advisory committee. This require-ment partly reflects the fact that gene therapy is novel, partly also the feeling that tampering with genes is in some way suspect and ought to be monitored.

In one sense, there is nothing special about gene therapy. For instance, patients with haemophilia are treated with factor VIII, the protein they lack which is necessary for the blood to clot properly. A gene therapy for haemophilia, which is under development, would give the patient the gene for factor VIII so they can make the protein themselves, rather than have it supplied from outside. Gene therapy is therefore simply transferring the source of the conventional drug from the outside to the inside of the patient.

It is too soon to say whether there are any special dangers associated with gene therapy. One remote possibility is that the new gene could insert itself into the genome in the 'wrong' place. That is it could activate an oncogene, or even inactivate some vital gene. So, to try to avoid this danger, experiments are going on to control where in the genome any new genes are likely to locate themselves.

The novelty of gene therapy has tended to eclipse the other contribution that genetic engineering has made to medicine – the production of recombinant human proteins. Some of these have been discussed already in Chapters 5 and 6. They include insulin, growth hormone, interferon and other immune system proteins, as well as various blood clotting proteins. Until the recombinant proteins were available, we had to use proteins extracted from human or other animal tissue. This always carried the risk of impurity or infection, so the recombinant versions are safer replacements.

Some of the proteins such as tissue plasminogen activator (tPA) may also represent a real therapeutic advance. tPA is the latest in a group of 'clotbusting' drugs that dissolve the clots blocking arteries to the heart that lead to heart attack. The problem with clotbusters is that they may lead to bleeding problems in other parts of the body. The latest evidence suggest that tPA is more effective than other clotbusters in improving survival rates after heart attack, if given with other anticoagulants. And with recombinant proteins there is always the possibility of protein engineering (see p. 111). This technology is being used to make tPA with fewer side-effects, such as unwanted bleeding, with the aim of producing a safer drug.

PART III

Biotechnology

9

The wide world of biotechnology

Biotechnology, the exploitation of biological materials and processes for human needs, has a long history. The ancient Egyptians applied mouldy bread to infected wounds for its antibiotic effect – today we turn that mould into penicillin. Also, the fermentation of fruits and grains to make wine, beer and spirits has been going on all over the world for thousands of years.

In biotechnology, we use microbes and cells as factories, and enzymes as the workers. Between them they turn out food, fuel, medicine and a wide range of other products in everyday use, with a market value of billions of dollars. Now genetic engineering and other techniques from molecular biology have given biotechnology a huge boost.

Enzymes are the master molecules of biotechnology

Much of biotechnology relies on the transforming power of enzymes. As we saw in Chapter 1, enzymes are proteins originating in cells, and each is usually specific for a particular biochemical process. Some enzymes are involved in basic cellular functions, such as extracting energy from food and making DNA. Others carry out more specialist tasks, manufacturing molecules that do not appear to be essential to the organism's survival. Instead they are used in 'chemical warfare', either to prevent competition or avoid predation. This is where antibiotics come from. We can imagine how, in ancient soils, competing communities would fight for their territory by using their own individual antibiotic molecules to try to wipe each other out. These products – known as secondary metabolites (metabolism is the term given to the enzyme-controlled chemical activity of the cell) – appear to be exclusive to the cells of microbes, simple marine organisms and plants. To date, no-one has found secondary metabolites in higher animals.

The enzymes that carry out general housekeeping duties in a microbial cell make useful products too. When a team of enzymes dismantles food molecules, such as glucose, to extract biochemical energy, a range of interesting waste materials is produced, especially if the process is carried out in the absence of air. These processes are generally known as fermentations, and ethanol (alcohol) is one of the most important fermentation products.

As well as being the basis of beers, wines and spirits, ethanol is an important industrial solvent. Before the rise of the petroleum industry in the 1920s, fermentation provided the main feedstock for the chemical industry. Fermentation products can be used as the starting point for the manufacture of textiles, rubber, explosives, biodegradable plastics, soap and many other materials.

The sterile world of the fermenter

The fermenter is the 'factory' in biotechnology – a vessel in which cells and microbes grow while the enzymes inside them turn out the desired product. Large-scale genetic engineering, such as the manufacture of recombinant insulin or chymosin, also takes place in a fermenter. In fact genetic engineering would have made no progress were it not for the skills of the traditional biotechnologists, who know how to run a good fermentation.

These scientists have a fanatical devotion to cleanliness and hygiene. They make sure that everything that comes into contact with fermenting cells or microbes is sterile; that is, free from contamination by other microbes. This usually means scrupulous attention to the sterilisation of instruments, apparatus, solutions, and chemicals with superheated steam.

Doing this on a laboratory scale is bad enough; having to do it on a commercial scale is even more demanding. A commercial fermentation may take place in a vessel with a volume of millions of litres. The microbes growing inside it need food, which is usually supplied as a soup with many different ingredients. They also, usually, need air or oxygen to extract the energy from the food. The vessel, nutrient soup and oxygen must all be sterilised too. And what about the biotechnologists operating the fermenter? They must be kitted out in helmets, protective suits, face masks and boots. We are all crawling with billions of microbes, and none of them must get an opportunity to try out life in the fermenter instead of on their biotechnologist host.

Any failure of sterilisation could mean the introduction of unwanted microbes into the fermentation. This can have several adverse consequences. The invading species may grow faster under the conditions of fermentation than the desired microbe does. If the contaminating microbe takes over the fermentation in this way, it is likely that little or none of the intended product will be made. The unwanted microbe might even be a pathogen, which would make the product dangerous to consume or use. Many such microbes produce toxins. Even if no toxin is produced, contaminating microbes can still affect the quality of a product – by generating 'off' flavours, for instance – or its yield, by consuming food meant for the producing microbial strain.

The dangers of contamination were highlighted when people being treated with the amino acid tryptophan began to fall ill with a mysterious illness. Tryptophan is used in the treatment of insomnia, depression, stress and premenstrual syndrome. It is very easy to produce by fermentation because so many microbes manufacture it in quantity as part of their general housekeeping duties (they need it to build into their proteins). By the end of 1990 over 1500 people in the US alone had reported disabling fatigue, muscle weakness, and inflammation of major organs such as the heart and liver, linked to taking tryptophan-based drugs. Some sufferers died. The illness was traced to a batch of the amino acid produced by a biotechnology firm in Japan. They had decided to change the microbial strain producing the tryptophan. The problem was that the new strain also produced a contaminant that was responsible for the adverse symptoms. Many countries withdrew tryptophan immediately, and the market for the product has still not recovered fully.

The Japanese scientists were unlucky with their choice of a tryptophan-producing microbe. There are around 6000 species of microbe that have been discovered, described and named. Estimates vary, but this is probably between 3 and 27 per cent of the total on the Earth today. Selecting the right species or strain (a variant within a species) for a particular fermentation is quite a challenge. Laboratory tests show which gives the highest yield of a product, but there is always the possibility of improving on nature. If a microbial population is treated with a mutagen such as X-rays or ultraviolet light, the rate at which the microbial DNA alters (or mutates) is increased (remember how Delbrück and his contemporaries used X-rays to study mutation, as described in Chapter 2). Occasionally the mutation hits a gene involved in the manufacture of the product, and alters it in such a way as to increase production dramatically. For instance it could affect a control region of the gene in such a way as to increase its expression. Experiments like this have led to ten-fold increases in production of products such as antibiotics. Genetic engineering comes in if you mix and match genes from different species, for example, genes from a high producer put into a microbe that grows fast.

The best conditions for growing the selected microbe for maximum production also have to be assessed. The guiding principle here is to find a strain that will accept a low cost diet, preferably one based on the waste material from another industry, such as molasses from sugar beet processing or whey from the dairy industry.

Once a growing microbial culture has been introduced into a sterile fermenter with the appropriate nutrient broth, the biotechnology team has to monitor conditions inside the fermenter carefully. Nowadays measurement of temperature, acidity, oxygen levels and various other factors are mainly computer controlled.

When the fermentation is complete, various processing operations are carried out to obtain the product from the broth. Even then, the spent microbes can be put to use – as yeast extract, for instance, which is a nutritious addition to the human diet, or as animal feed.

Biotechnology in medicine

Drugs

Traditionally, secondary metabolites such as antibiotics have been the main drugs to come from biotechnology. Now, however, the production of vaccines and antibodies is increasing in importance, thanks to the application of new advances in cell biology and genetic engineering.

Humans have derived enormous medical benefits from secondary metabolites, whether prepared by native healers in a rainforest or by advanced fermentation techniques. The biggest lifesaver is probably penicillin. Today we take antibiotics for granted, and it is hard to believe that less than 100 years ago tuberculosis, diphtheria or even an infected cut could be killers. Alexander Fleming's accidental discovery of penicillin – the compound manufactured by a common mould – went unexploited for many years. Even when it was finally taken up just before the Second World War by Ernst Chain and his colleagues in Oxford, the first patient to be treated with it died of overwhelming infection when the supplies of the drug ran out. War provided the stimulus for full-scale production, and soon penicillin-producing microbes were being grown in trays, dishes, and other small fermenting vessels at sites all over Britain. Penicillin is still made by growing *Penicillium crysogenum*, although today giant fermenters are used.

Microbes are undoubtedly valuable resources, but many more medicines are derived from plants. These range from established drugs such as aspirin to those like taxol, whose potential is still being evaluated. Indeed, new uses for aspirin are being found all the time – from thinning blood to possibly preventing Alzheimer's disease.

Taxol is a particularly interesting example of how the teams of enzymes working in the cellular laboratories of plants often surpass the best efforts of the world's multi-billion pound pharmaceutical industry. In the early 1960s, the US National Cancer Institute (NCI) launched a drive to find new anti-cancer drugs in plants. Taxol comes from the bark of the Pacific yew and was discovered in Oregon in 1962. It turns out to be the most important of the 110000 plant compounds tested for anti-cancer activity by the NCI team between 1960 and 1981.

Taxol is particularly effective for patients suffering from advanced breast and ovarian cancers. It is also used to treat leukaemia, lung cancer and melanoma. It stops cancer cells assembling their cytoskeletons properly. The cytoskeleton is a support system built of protein fibres that is essential for cell division, mechanical support and other functions. Without a cytoskeleton, the cancer cells are effectively paralysed and die. Recent reports suggest that taxol might also be an effective treatment for polycystic kidney disease, which accounts for ten per cent of the demand for kidney dialysis treatment and transplantation.

Taxol could be one of the most useful drugs of the decade. But there is one drawback. The bark of a whole tree produces only one dose of taxol. To treat 500 patients, 3000 trees have to be sacrificed. Already environmentalists are protesting that the forest habitat of the spotted owl is threatened by taxol production.

The chemical structure of taxol was discovered in 1971. It is a small, but complicated, molecule. Chemists round the world took up the challenge to synthesise it in their laboratories, and spare the yew trees. After 23 years of effort, Kyriacos Nicolau and his team at the Scripps Research Institute have finally found a way of making taxol – but no-one knows yet if the synthesis can be made cost-effective. A compromise would be to let the enzymes carry on making taxol, but to spare the trees by removing a sample of taxol-producing cells to grow in a fermenter.

Vaccines

Hepatitis and AIDS are both major diseases which are caused by viruses. There are no drugs that have been as successful against viral diseases as antibiotics have been against bacteria. Instead we have relied on the protection given by vaccines. Thanks to vaccines, smallpox has been eradicated, and polio and measles should soon be wiped out too. But major problems with viral infections remain. Advances in biotechnology are now leading to the production of more effective vaccines.

Infectious microbes carry substances called antigens on their surfaces that stimulate the body's immune system (there is a great deal more to it than this, but for the purposes of this discussion we need only the absolute basics). The main feature of this response is the production of proteins called neutralising antibodies, which destroy the invading microbes. The problem is that these defences are often assembled too late to ward off the damage being done by the invader. Ideally, one would like the neutralising antibodies to be present at the outset. This is what vaccination does.

A vaccine is a 'presentation' of the antigen of a microbe to the body that stimulates the production of the appropriate antibodies. These stay in the blood – usually for years – so that when the real microbe comes along they are ready to tackle it straight away.

Edward Jenner was the first to try out a vaccine in 1796. Cowpox is a viral disease that is far milder than smallpox, and Jenner had noticed that milkmaids who had had cowpox did not get smallpox. He thought the cowpox infection gave them some protection and made the first crude vaccine of fluid containing the cowpox virus. He injected this into a young boy, using a thorn as a syringe, and a few weeks later daringly injected him with smallpox! Fortunately, the cowpox virus had stimulated the production of antibodies, which then attacked the smallpox virus.

Today similar principles are used in developing vaccines. If a live vaccine such as Jenner's is used, it obviously has to be a weakened (attenuated) strain. Live vaccines score by replicating in the body and giving a high concentration of antibodies, but there is always the worry that the weakened strain could mutate to a virulent strain while it lives.

An alternative is to use dead virus. Again, these can generate an effective response, but there is a danger that a few organisms might survive the heat or chemical treatment used to kill the virus. The third type of vaccine isolates just the antigen from the virus. Obviously this is safe, as whole viruses are not being used. The challenge is to use the right antigen: viruses may have several molecules on their surface that could act as antigens.

All three approaches require the virus to be grown in a fermenter. Viruses are not able to grow on their own; they need to first invade a cell. Animal cells, such as ovary cells from insects or Chinese hamsters, are used for this. Inevitably extra care has to be taken around fermenters where live viruses are being grown. Each batch of killed virus or antigen has to be scrupulously tested for the presence of live virus before it can be packed.

Genetic engineering has provided a way of getting pure antigen without having to use live virus. The first genetically engineered vaccine was for human hepatitis B. This is a severe, and potentially fatal, infection of the liver. Having hepatitis B greatly increases the chances of later contracting liver cancer. The vaccine is made by transferring the gene for a carefully selected antigen, from the surface of the hepatitis B virus into yeast. The yeast multiply in a fermenter, making large amounts of the antigen as they do so. This can then be isolated and used directly as a vaccine.

There are many similar approaches. For example, you can splice together viruses from plants (which cannot harm humans) and antigens from human viruses. Infecting plants such as cowpeas with these viruses provides a way of multiplying them, so that supplies of vaccines can be harvested a few weeks later. Vaccines against HIV and animal diseases such as foot and mouth disease and mastitis are being made in this way. Or you can turn to chemical synthesis to make large amounts of pure antigen. Usually the antigens are proteins, and often just part of that protein, a segment known as a peptide, will be enough to get the required immune response. A peptide synthesiser, which is a machine that simply strings together amino acids to make a peptide, has produced an effective vaccine against malaria. Although malaria is caused by a protozoan, not a virus, the principles of vaccine development still apply.

Antibodies

The antibodies produced to an antigen, as described above, are not formed from one single protein. Instead they are a mixture of proteins produced by white blood cells, each responding to a different part of the antigen molecule (these parts are called epitopes, and usually they consist of just a few amino acids). These so-called polyclonal antibodies have little potential for clinical or diagnostic use, because we cannot be sure which bit of antigen they are directed against. However, in a cancer called multiple myeloma, one type of white cell begins to divide uncontrollably, producing large amounts of just one kind of antibody. Antibodies like this are called monoclonal antibodies (MAbs). They lock on to a specific epitope. In contrast to polyclonal antibodies, MAbs have enormous clinical potential. They can be used to seek out, and bind to, any protein in the body. So they could be used to neutralise an abnormal cancer protein, for example, or to pick up proteins associated with infection in blood samples in a diagnostic test.

The technology for making MAbs was developed by Georges Köhler and César Milstein, working in Cambridge in 1975. They found that if you fused together white blood cells from a mouse that had been injected with an antigen and myeloma cells it was possible to get hybrid cells, called hybridomas, that would turn out large quantities of MAb to the antigen. Hybridomas can be cultured in a fermenter like any other cell. So many biotechnology companies today are busy turning out MAbs to many different antigens.

MAbs are already being used for a variety of clinical applications, such as to deliver drugs to cancer cells, by 'recognising' and homing in on antigens on the cancer cells' surface. Once the MAb and cell have 'met' the antibody can deliver its deadly payload of drug, leaving normal cells unaffected. The same idea can be used for more precise imaging of tumours prior to surgery or treatment. Here the MAb can be tagged with a dye or radioactive flag, so that the tumour can be 'outlined' on imaging apparatus.

MAbs are also extremely useful in the purification of proteins from blood plasma. They lock on to the desired protein – and nothing else. After the rest of the blood has been separated from the MAb–protein pair, the purified protein can be recovered by chemical treatment.

MAbs produced in the traditional way, as described, have the drawback that they are mouse proteins because the white blood cells come from experimental animals. So they are not ideal for human clinical use. Greg Winter and his team at the Laboratory for Molecular Biology in Cambridge are now adopting the MAb technology to produced 'humanised' antibodies.

Humanised MAbs can be used clinically to 'neutralise' undesirable protein molecules. For instance, some of the proteins of the immune system are present at an abnormally high level after organ transplantation or in inflammatory conditions such as arthritis or septic shock. Once an MAb homes in on them, they are effectively put out of action – exactly as if they were an infectious microbe being attacked by the immune system.

A taste of enzymes

Biotechnology in the food industry relies on the power of enzymes to transform raw materials such as milk, grains and fruit into tasty and interesting products. The enzymes used are usually obtained from fungi, and have been certified safe by the regulatory bodies. Of course, enzymes have been producing food and drink long before anyone realised they existed.

The dairy industry uses enzymes in a wide range of bacteria and fungi to make yoghurts and cheeses. Cheese-making depends on the action of protease enzymes found in the stomachs of certain animals. This explains why nomadic tribes of Eastern Europe and Western Asia sometimes made cheese by accident when they carried milk in bags made from animals' stomachs. The most effective of these enzymes for cheese-making is chymosin, the main constituent of rennet, which is an extract from the stomach of calves.

Milk contains soluble particles of a protein called casein. The solubility comes from the short chains of sugars and amino acids called glycopeptides on the surface of these particles. Added chymosin detaches the glycopeptides so that the particles clump together. The liquid milk turns into solid curds and liquid whey. The curds are compressed and matured to make cheese. It is bacteria and fungi in the milk that produce characteristic cheese flavours and aromas during the maturing process. Colours – such as the blue veins in Stilton – bubbles and holes, and rinds are also the products of microbial action.

The demand for vegetarian cheese was originally met by trying to use proteases from plants rather than animals. But the results were disappointing; it seems there is something rather special about calf chymosin. Genetic engineering has been able to provide a solution – vegetarian cheese with all the flavour and texture of the traditional cheese – as we saw in Chapter 5. Genetically engineered chymosin must surely be just the beginning, so great is the scope for gene manipulation in the dairy industry.

The fruit juice and wine industries are heavily dependent on another enzyme, pectinase. This breaks down pectin, which is a sort of biological glue that keeps the cells together in the flesh of various fruits. Pectin also makes jam solidify, but it stops the juice running out of fruit and causes cloudiness in wine. The judicious use of pectinase in juice extraction allows the flow of millions of tonnes of fruit juice a year.

Enzymes also help to control the appearance and flavour of fruit juices. We are accustomed to cloudy orange juice and clear grape juice, while apple juice is either clear or cloudy. When juices are first extracted from the fruit, they often contain suspended particles of protein or starch. The presence of residual pectin makes the juice viscous and difficult to clarify by simple filtration. Adding pectinase after extraction thins the juice and also breaks down the pectin that coats the surface of the suspended particles. Once stripped of this coat, the particles clump together and float to the bottom of the juice, where they settle.

The brilliant colour of juices from blackcurrants, cranberries and grapes comes from chemicals in the skins of the fruit. The release of these substances into the juice is aided by the use of cellulase, which breaks down the cellulose in the skins during processing.

Finally, the taste of fruit juices can be controlled by the addition of enzymes. For instance, grapefruit works well as a first course because it contains a compound naringin, which stimulates the taste buds in such a way as to enhance the food to come. However, a little naringin goes a long way. Too much makes grapefruit unbearably bitter, and manufacturers try to overcome this by blending sweet and sharp varieties of the fruit. Alternatively, the enzyme naringinase can be added to grapefruit juice to control the bitterness, because it breaks down naringin to a non-bitter chemical.

All the above processes are a key part of commercial and home wine- and cider-making. Yeast is added to the juice or fruit pulp (or wild yeasts on the skins of the fruit may be used alone). A set of enzymes in the yeast cells sets to work on the sugars in the fruit and extracts biochemical energy from them. Alcohol is merely the waste product of this process. Pectinase is also useful for clearing hazes from country wines such as that made from apples.

Brewing beer relies on the enzymic breakdown of starch to make glucose, which is then fermented. The raw material of beer is barley grain, and the brewer taps into the natural process by which the barley converts its stores of starch into glucose. By allowing barley to germinate – a process called malting – high levels of enzymes called amylases are produced in the sprouting grain. These snip off individual glucose molecules and short chains of glucose molecules from the gigantic starch polymers for immediate use by the developing barley plant. Brewers sometimes cut short the germination process and supplement the malted grains with amylases from fungi to stop the barley embryo consuming too much of the glucose, which could be fermented into beer.

The organism that turns the glucose into alcohol is the yeast *Saccharomyces cerevisiae*. Over the years many new strains of *S. cerevisiae* have been cultivated. The problem with most of the traditional strains is that they cannot ferment the longer glucose chains that come from starch – molecules known as dextrins. A related strain called *Saccharomyces diastaticus* can ferment dextrins, but gives an unpleasant-tasting beer. However, the gene that is responsible for dextrin breakdown has been transferred from *S. diastaticus* to *S. cerevisiae*. The result is a low carbohydrate beer, which tastes good too.

Brewing and baking have much in common. Both rely on the conversion of starch into glucose, which is then fermented by a yeast. However, the raw material for bread is wheat grain. Again, the grain's natural amylases may be supplemented by the addition of fungal enzymes. Fermentation in baking does not produce a great deal of alcohol – what is produced contributes to the characteristic smell and flavour of a fresh loaf. However, the other waste product of fermentation, carbon dioxide, makes the bread rise when it is baked in the oven.

Biotechnologists have long speculated on the possibility of making wine and beer from old newspapers. The problem is that in newspapers the polymer is cellulose, while in grains and fruits it is starch. They are both made of glucose units, but they are linked together differently. Amylase cannot break down cellulose, but cellulase can. However, the product is not easily fermented. To date a low grade alcohol that can be used as a fuel supplement has been produced from waste paper. But we will have to wait a few years before that crate of old newspapers can be turned into a crate of the finest champagne!

Enzymes in the wash

Biological detergents contain tiny amounts of enzymes that break down stains. Otto Röhm launched the first biological detergent in 1913. This was an extract of pig pancreas that contained enzymes of the protease family. Protease break down proteins, and Röhm argued that the dirt that builds up in clothes is composed of human fats and proteins. The protein component of this makes the dirt adhere to the fabric. A protease enzyme can chop up the proteins into amino acids, so that the dirt dissolves in the washing powder.

However, the first biological detergents were crude and ineffective because the enzyme molecules tended to unravel in the alkaline conditions of a typical wash. The search was on for a protease that was compatible with alkali, and in the 1960s the Danish company Novo Industri (now Novo Nordisk) found one, in the bacterium *Bacillus licheniformis*. The company went on to develop detergents containing these enzymes, with the aim of removing stains and smells from the heavily soiled overalls of people working in the fishing and meat industries.

In recent years, other microbial enzymes have been added to detergents. Amylase breaks down starchy stains such as tinned spaghetti, gravy and baby food, while lipases dissolve fat-based stains including lipstick and grease. The most recent development has led to detergents that condition the material itself. Cotton-based fabrics are based on cellulose, the world's most abundant natural product. Cellulose molecules are long strings of glucose molecules that group together to form the fibres characteristic of cotton. The action of washing can damage the cellulose fibres.

Fibre fragments then form a characteristic 'fuzz' on the surface of the material. If the enzyme cellulase is added to a detergent, it can digest these fragments, improving the feel of the garment.

The advantage of enzymes in detergents is that they allow the wash to be carried out at a lower temperature than a traditional wash. Proteins are delicate molecules that depend on the precise maintenance of their shape for their function. Temperatures much above 50 °C will cause the carefully folded protein to unravel or denature. This is why a hot wash is not a good idea for removing blood or egg from a garment. The high temperature merely degrades these protein-based stains, making them cling more tightly to the fabric. This is because atoms in the unravelled (degraded) protein become free to make chemical bonds to atoms in the fabric. The enzymes in biological detergents, which are themselves proteins, do not survive a hot wash and work best at moderate temperatures.

Cooler washes save energy and, because enzymes are directed towards specific stains, should give a better result than soap or detergent alone. They also fit in well with the trend towards phosphate-free and liquid detergents. Currently, biological detergents account for over 80 per cent of the market in Western Europe.

However, there are a couple of drawbacks to the use of biological detergents. First, as enzymes are foreign proteins, people exposed to them can suffer an allergic response. This happens if the immune system, which is tuned to respond to harmful foreign proteins such as those carried on the surfaces of some microbes, is hypersensitive and starts to respond to proteins that are not, in themselves, harmful. Some workers in detergent factories did experience severe allergies from exposure to enzyme dust in the late 1960s. The resulting bad publicity caused sales to plummet and the companies began to put more emphasis on safe handling procedures. Nowadays detergent enzymes are packaged so that direct exposure to them is very unlikely.

Second, silk and wool are natural materials made of a protein called keratin (the same protein that is a major component of hair, skin and nails). So soaking these fabrics in biological detergents might change them, because the proteases could break down the keratin.

10
Plant power

Plants support life on earth, primarily by transforming the energy of the Sun into the chemical energy of food during photosynthesis. They are also the raw materials for medicine and natural textiles.

For the last several thousand years, we have done our best to mould plants to our needs with agriculture. Traditional breeding methods have tried to exploit the best of what the plant world has to offer. Now biotechnology and genetic engineering are opening up major new possibilities, not only of breeding plants with desirable characteristics but also of creating entirely new plant species. To date, the commercial success of plant biotechnology has been limited. However, its scope is potentially very broad – from providing food for a growing world population to introducing novel flowers such as blue roses and brick-red petunias.

One cell, one plant

Biotechnology applied to plants takes advantage of the fact that – unlike animal cells – plant cells are totipotent. This means that the pattern of gene expression in a plant cell gives it the potential of becoming any kind of cell in the mature plant. So a plant cell could end up in the stem, leaf, flower or root. Animal cells, however, are pluripotent. They can only become one of a limited group of cell types. Stem cells, for example, which are produced by the bone marrow of humans and other vertebrates, could become any one of a number of white blood cells used by the body to fight infection. But they could never become nerve cells, or muscle cells. So although both animal and plant cells contain a complete copy of the DNA needed for every type of cell in the mature organism, the pattern of gene expression in each – that is, the way some genes are turned on or off at a certain time – means that their developmental paths are very different.

In theory then, you can get a complete plant from a single plant cell. Keen gardeners already know this. When they take a cutting and root it, with luck and experience a whole plant can be obtained from a leaf, or a stem, of an admired plant from a friend's garden. (But there is no chance of taking a hair from the same friend's pet poodle and having it grow into a dog!) Where a farmer's livelihood or a company's investment is at stake, this process takes place under controlled laboratory and field conditions. Creation of a set of genetically identical plants, known as clones, from cells of a plant with desirable characteristics, is one of the key commercial activities of plant biotechnology today.

The success of these enterprises depends on a technique called tissue culture, which was developed during the first few decades of the twentieth century by biologists who were fascinated to discover that cells could stay alive outside an organism. Inevitably this led to much discussion about the relationship between cells and organisms. A myth grew up that cells were immortal, and it was being part of an organism that made them die. As we saw in Chapter 8, this is only true of abnormal cells, such as cancer cells. Normal cells have a limited lifespan, but they can certainly exist in their own right as independent entities, much as single-celled organisms like bacteria do.

Tissue culture was the practical spin off from this. Cells from both animals and plants were grown in glass flasks containing a sort of soup that had all the ingredients necessary to provide a healthy diet for growing cells. The cells can come from a tiny piece of tissue taken from the organism and then digested with an enzyme to give separated cells. The whole tissue can also be grown in culture. Scientists at the Strangeways Laboratory in Cambridge, led by Honor Fell, one of the great pioneers of tissue culture, showed that even small organs – such as limbs from chicken embryos – could be grown in culture too.

Here is how a plant tissue culture programme works. Suppose

your local natural history society is concerned about the last remaining rare orchid in a nearby wood. With tissue culture, the society could clone the orchid, and produces several hundred more – given the resources and facilities. First a small tissue sample should be taken from the orchid – with a sterile scalpel. Although any part of a plant can be used to start off tissue culture, in practice the meristem, which is the actively growing tip of a root or shoot, is favoured. This is because the meristem is most likely to be free from virus, fungus or bacterial infestation, which is vital when disease-free clones, perhaps from a certified plant collection overseas, are being created.

The orchid tissue would then be sterilised simply by placing it in a dish of dilute bleach solution for five minutes or so. The next step is to place it in a flask or dish containing a nutrient solution. The composition of this chemical cocktail is very finely tuned and consists of sugars, vitamins, and other factors whose role in cell growth remains obscure yet vital. The development of such a medium may take years and owes as much to culinary intuition as it does to hard science.

In this hospitable environment, a mass of undifferentiated orchid tissue called callus should appear after a few days. Callus looks rather like apple sauce, surrounding the original tissue sample. Tiny samples of callus can be removed from the original vessel with sterile tweezers and placed in a new nutrient vessel to grow on.

Then the clumps of callus are transferred into a new culture solution – one that contains a carefully designed mixture of plant hormones to encourage development of roots and shoots. Hormones are chemical messengers that stimulate cells to carry out certain biochemical activities at certain times. For instance, human sex hormones orchestrate the menstrual cycle, and the development of the sexual organs. In plants, hormones control not only the formation of roots and shoots but also the ripening and the dropping of fruit. The study and exploitation of plant hormones has been of enormous importance in horticulture and agriculture. Hormones are used by gardeners to promote rooting of cuttings, and by farmers to make fruit drop from trees in an orchard at the same time for a speedier harvest.

One important ingredient of the tissue culture hormone mixture is 2,4-D, which encourages shoot formation. 2,4-D is also used as a weedkiller. It forces weeds to grow uncontrollably so that they eventually weaken and die. No-one wants this to happen to the vulnerable plantlets that will emerge from the tissue culture programme, so only tiny amounts of hormones are used to make the callus develop into a plant. If all goes well, you should have lots of orchid plantlets – the exact number being dependent on the scale of the operation – ready for planting out and growing on within a few weeks.

Besides the preservation of rare plant species there are several other reasons for using plant tissue culture. It is faster than waiting for some slow-growing species, such as trees, to produce seed. It can also save crop devastation by disease if a known disease-free strain is cloned. Of course it also sidesteps the potential pitfalls of plant breeding where genes are mixed. With tissue culture you know exactly what you are getting (with one exception, which will be considered later), so you can produce more of a high-yielding, rare, or otherwise desirable plant.

The process of producing a whole plant from a lump of callus tissue is known as regeneration. Many plants have already been regenerated from callus. These range from fruits and vegetables to trees and ornamental plants such as orchids. There is no scientific reason why all plants should not be grown and regenerated by tissue culture once the right system has been developed. However, with the exception of rice, the cereal plants – which are mono-cotyledons (or monocots) because they produce only one seed leaf – have proved hard to grow in tissue culture. Dicotyledons (or dicots), which are the broad-leaved plants, are far easier to regen-erate. Tobacco may be devastating to human health, but it is the biotechnologist's friend because it is so amenable to regeneration. Along with its close relations tomato and potato, tobacco is the workhorse of many research programmes – just as *E. coli* and the fruit fly *Drosophila* are extensively used as model organisms in other areas of genetics.

Tissue culture can be a fairly low-tech operation, and is there-fore usable by a wide range of growers, from small farmers to giant multinationals. So in North Vietnam farmers have used tissue culture to overcome problems in the supply of European seed potatoes caused by the Vietnam war. Using tissue from the International Potato Centre in Peru, the farmers and their families regenerated three million disease-free potato plants, which now give them a steady income. They use discarded US gas cylinders as steam sterilisers, and banana leaf pots to root the plantlets.

At the other end of the investment scale the chemical giant Unilever is building huge plantations of high yielding, cloned oil palms in Malaysia. Palm oil is an important crop. It is used as a cooking oil and is high in chemicals called free-radical scavengers that may reduce the risk of cancer. There are also many industrial uses for palm oil – from plastics to soap. Unilever hopes that the plantations, which could contain up to one million trees, will have four to five times the yield of oil compared to a plantation produced by traditional breeding.

Tissue culture can also be used as a sort of fermentation process, with the emphasis on what the plant can produce, rather than the plant itself. Chemicals produced by plants are called phytochemicals and, as was discussed in Chapter 9, plants and microbes are capable of some elaborate chemistry. Phytochemicals can be used as drugs, dyes, perfumes and insecticides, and of course there are also major cash crops such as coffee, tea and cocoa.

But plant tissue culture has not, so far, been very successful as a source of phytochemicals. The cells tend to clump together and form plants, rather than acting as individualised production units for whatever chemical they specialise in. The best hope appears to lie in the cultivation of hairy root cells. Japanese scientists found that they could get large amounts of the red dye shikonin from the root cells of the wild herb *Lithospermum erthyrorhizon*. Shikonin, which is used in the Japanese flag, and to colour lipstick, is traditionally extracted from the root of the plant. If the cell culture succeeds commercially, then the herb gatherers will find themselves out of a job. Here in Britain, hairy root cell culture is beginning to yield large amounts of tropane alkaloids, chemicals that are extracted from plants such as deadly nightshade and are widely used as anaesthetics.

New genes, new plants

Plant cells in tissue culture are good candidates for genetic engineering. The insertion of foreign DNA into a plant cell is called transformation. If this is followed by successful regeneration then the result is a transgenic plant (Fig. 10.1).

The problem with making plant cells take up new DNA is that they contain tough cell walls made of cellulose, a long stringy molecule. These make the plant cells a little like bricks – good for building the plant, giving it structure and support, but impenetrable to microinjection techniques for introducing DNA into the cell.

Nevertheless there are ways through the plant cell wall. One approach is to have the new DNA hitch a ride on a bacterium called *Agrobacterium tumifaciens*, which preys on broad-leaved plants, infecting them in the way another bacterium might give you a sore throat or upset stomach. *Agrobacterium tumifaciens* induces a tumour called a crown gall on the plant. The bacterium contains a plasmid that bears the genes for creating the crown gall. The trick is to incorporate the foreign DNA into this plasmid, so that *Agrobacterium* can smuggle it into the plant.

This is done by using enzymes to cut, trim and splice the plasmid DNA – with techniques similar to those described in Chapter 5. If all goes well, *Agrobacterium* can be converted into an excellent vehicle for transporting foreign DNA into plant cells. Early experiments with tobacco, petunia and carrot plants showed the power of the *Agrobacterium* route.

Unfortunately *Agrobacterium* does not infect monocots – which are exactly the plants we are interested in for food production. However, there is a way for DNA to penetrate cereals. The solution is simple: strip away the cellulose cell wall with enzymes to leave 'naked' cells called protoplasts. The enzymes used – cellulases and pectinases – are those used in detergents and fruit juice processing. They digest the components of the cell wall, making the cell far more porous to DNA.

Then there are two ways of introducing the new DNA into the protoplast. Electroporation, a form of electric shock treatment,

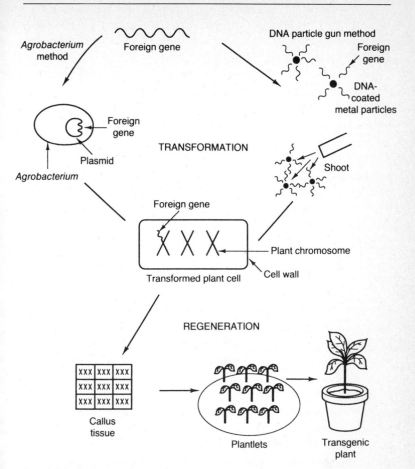

Fig. 10.1. Creation of transgenic plants by genetic engineering. A foreign gene – for a property such as pest resistance, maybe – is inserted into plant cells, using a technique tailored to the individual plant. For some, such as tomato and potato, a bacterium called *Agrobacterium tumifaciens* smuggles in the gene. Others, such as cereal crops, are more susceptible to direct methods of gene transfer such as 'shooting' with DNA-coated metal particles. This part of the process, which results in plant cells with inserted genes, is called transformation. The transformed cells are then grown into a mass of tissue called callus. Addition of hormones to the culture medium produces roots and shoots, turning callus into plantlets that can be grown up into mature transgenic plants. Turning plant cells into plants is known as regeneration.

punches tiny holes in the cell membrane. The new DNA sneaks into the cell through these. Later, the holes close up again, leaving the cell undamaged. In a more recent development, protoplasts are 'shot' at with a 'gun'. This contains tiny gold or tungsten projectiles coated with the DNA to be introduced into the plant cell.

Transformed plant cells are selected in the same way as recombinant bacteria. That is, marker genes are transferred along with the gene of interest. Generally these are for antibiotic resistance – the kanamycin resistance gene is a particular favourite. So plant cells that have the required gene will be resistant to kanamycin and can be clearly distinguished from plant cells that do not have the gene: the antibiotic will kill these.

Engineering the world's food supply

We now have the technologies to genetically engineer any plant. However, the prospect of creating a generation of designer superplants is a long way off, mainly because some of the most interesting attributes of plants remain poorly understood at the genetic level. Radical changes in plants, via genetic engineering, will have to wait until more of their basic biology is worked out.

The two main things that plants can do – and animals cannot – are photosynthesise and fix nitrogen (note that some bacteria photosynthesise, and there is no nitrogen fixation without bacteria). Both processes have a major impact on our food supply, so biotechnologists would love to have better control of them through genetic engineering.

At a recent conference the world's top plant biotechnologists voted improvements in photosynthesis as having more than twice the commercial potential of anything else they could do to engineer plants. In photosynthesis, simple carbon dioxide molecules in the atmosphere are built up into more complicated ones that can be used as food. Glucose, which has six carbon atoms to carbon dioxide's one, is the first product of photosynthesis. Plants use glucose directly but also elaborate it into other substances such as starch, cellulose, fats and proteins. Building these molecules takes energy, which plants get from trapping sunlight using their green pigment, chlorophyll. The biochemical details of photosynthesis are fascinating, but complicated. Perhaps the most important thing to remember is that, as with all biochemical reactions, enzymes are involved at every step, building each molecule from its predecessor.

Photosynthesis has always driven life on Earth, right from the days of the ancient cyanobacteria nearly four billion years ago. All modern life forms depend on it too (except for some bacteria), either directly or indirectly. You can trace most of your daily diet back to photosynthesis, even if you never gather plant food directly.

The problem with photosynthesis is that it is really not very efficient. The blame can be laid at the door of one of the key photosynthetic enzymes, ribulose-1,5-bisphosphate carboxylase, known in chemical shorthand as Rubisco. Compared to other enzymes, Rubisco lacks tenacity when it comes to holding on to its substrate – in this case carbon dioxide. It also fails to devote all of its time to photosynthesis; sometimes it can be found helping out with another, competing, reaction.

Genetic engineers would like Rubisco to work harder. One way would be to rummage through the rich diversity of the plant world and find Rubiscos that are more efficient than the ones in important food plants (until now, we have discussed Rubisco as if it were just one enzyme; in fact Rubisco is more of a family name and each plant will have its own version). Or they could take a long look at exactly what features of Rubisco's three-dimensional structure control the way in which carbon dioxide is held by the enzyme. In situations like this, usually just a few amino acids are involved. Changing these – by the protein engineering techniques discussed in Chapter 5 – could transform the sluggish Rubiscos into dynamic hard-working new versions.

Nitrogen fixation is the other great biochemical achievement of plants. This is the transformation of nitrogen, which is the most abundant gas in our atmosphere, into ammonia. As nitrogen molecules consist of two nitrogen atoms, while ammonia molecules are one nitrogen atom surrounded by three hydrogens, this might not sound like much of an achievement. In fact it is very difficult to do – without enzymes. Making ammonia in industry requires the combination of nitrogen and hydrogen gas under conditions of high temperature and pressure, and even then a metal catalyst is essential to make the process economic.

In leguminous plants, such as peas, beans and lupins, nitrogen fixation occurs with the aid of certain soil bacteria, usually of the *Rhizobia* species. The action takes place in specialised organs, nodules, on the roots. The bacteria stimulate nodule formation. Then they invade these tiny compartments, and set about welding together nitrogen and hydrogen atoms, with the aid of a complex enzyme called nitrogenase, into ammonia molecules.

Nitrogenase is exquisitely sensitive to oxygen. Just a trace of the gas destroys it, and ammonia production grinds to a halt. This is why nitrogen fixation cannot occur in the soil. In the nodule, the plant is able to provide biochemical protection for the bacterial nitrogenase. It manufactures a protein called leghaemoglobin. As the name suggests, this is a form of haemoglobin. It grabs hold of any oxygen in the vicinity of the root nodule before it can destroy the nitrogenase.

Ammonia can readily be built into amino acids and proteins; this is why the plant needs it. Without nitrogen fixation, it would have to rely on the action of other soil bacteria, known as the denitrifiers. Alternatively, artificial fertilisers, such as ammonium phosphate and urea, could be used to boost its supplies.

So although it seems arrogant to think of interfering in the beautiful plant–bacteria symbiosis that is nitrogen fixation, the prospect of extending this facility to a wider range of plants is tempting. Yields could be boosted, and we could cut down on the use of fertilisers – with the accompanying problems of run-off into the water supply – at the same time.

One option is to encourage the development of symbiosis between other plant and bacterial pairs, by transferring genes responsible for nodule formation. A more radical solution would be to dispense with the need for symbiosis altogether and make the plant self-sufficient by giving it all the genes needed for nitrogen fixation. However, as there are 17 different bacterial genes involved in the process, this would be an extremely ambitious project! To date, no-one has experience of this kind of multi-gene transfer between organisms.

Manipulating photosynthesis and nitrogen fixation might begin to look a little more realistic when information from the various plant genome mapping projects has been analysed and understood. The land cress (*Arabidopsis thalania*) is the model organism for this work. It has a small genome with few DNA repeats. Its rapid generation time of six weeks means that results from genetic experiments are soon available. Scientists collaborating internationally on *A. thalania* mapping are already gathering and swopping new insights into how plants develop and flower, and how they resist environmental stresses such as cold weather and predators such as viruses, fungi and bacteria. So far, the genetic engineers have used what is known about how a plant functions to try to improve the chances of crop plants surviving attack by microbial predators and adverse environmental conditions to ensure healthier crops – and more food in the world's larders.

By the middle of the twenty-first century, the global population could reach nearly nine billion. Depending on the extent to which people still go hungry, existing farmland will have to provide between two and six times as much food (the latter figure is what would be needed if everyone had as much to eat as we do in the West). We could, of course, turn more land over to agriculture, but this might lead to environmental and political problems. Where biotechnology could come in is to boost yields, which are well below maximum because of pests, weeds, and climate problems such as drought.

There are two main approaches to increasing the food supply. Either we can stick to the crops currently grown, and try to improve them, or we can diversify and try out new food plants. The first option means concentrating on high yielding varieties, and finding ways of reducing losses. The so-called 'Green Revolution' of the 1950s and 1960s boosted wheat production by 100 per cent by the planting of more productive varieties and improved cultivation techniques. But there was comparatively little emphasis on selecting varieties with good disease resistance. Instead, there was an increasing reliance on chemical pesticides and herbicides to eradicate attack on the crops by predators and weeds.

Old-fashioned science textbooks boast of what chemistry had done for agriculture, citing the insecticide DDT as an example.

Nowadays DDT has a somewhat tarnished image: it is toxic to humans and wildlife and lingers in the soil for years. Many countries have banned its use and its place has been taken by less harmful compounds. But the demise of DDT may have led to the resurgence of the mosquito, and a continuing problem with mosquito-borne diseases such as malaria and dengue fever.

The other characteristic of the Green Revolution was to develop a reliance on relatively few plants for food. We now use only 20 species to obtain 90 per cent of our food supply, out of thousands of potential food plants. Cereals – flowering grasses whose seed is used for food – roots, and legumes dominate. Half of our calorie intake comes from wheat, rice and maize. Barley, yam, cassava, potato and soya account for most of the rest. Although we could try to increase the range of plants we use for food, which would represent a major shift away from the monocultures that are more vulnerable to disease, the approach adopted by the 'gene revolution' has so far been to improve existing crop plants.

The kind of basic, fundamental biological knowledge needed to develop ideal crop plants is being gathered from various cereal genome projects, which parallel the mapping of *Arabidopsis*. There are already detailed genetic maps available for barley, rye, millet and wheat. But the wheat genome, which has been extensively studied in Britain, is huge, containing far more DNA than the human genome. Wheat is hexaploid: its chromosomes are arranged in groups of six (three pairs of each). Altogether it has 42 chromosomes, each one having as much DNA as 8 human chromosomes or 25 rice chromosomes. The wheat genome is also stuffed with apparently useless repeat sequences – frustrating for scientists who just want to extract the genes.

However, Japanese scientists studying the much smaller rice genome have found that the order of the genes in rice and wheat is the same. This is a reminder of the fact that both plants have evolved from the same ancient grass and 60 million years ago they diverged from their common evolutionary pathway to become two different species. Now rice can be used as a master plan for other cereal genome work. Plant scientists round the world are now pooling information and hoping to speed up the location of agriculturally useful genes on major crop cereals.

Wheat, in 1992, was the last of the major crop plants to join the transgenic club. This opens up the exciting prospect of a global agriculture based upon a precisely honed genetic fitness. However, it will be many years before transgenic plants are available to farmers on a commercial basis – although numerous field trials have been carried out. The first generations of these are likely to be plants engineered for resistance to disease, insect attack and herbicides.

All gardeners have first-hand experience of plant diseases and predators. Greenhouse tomatoes are devastated by viruses, mould attacks strawberries, and of course the lives of slugs are dedicated to eating your favourite vegetables and flowers. It is annoying (perhaps even heartbreaking) to see what pests can do in the allotment or garden. Imagine this multiplied on a farm or plantation to get some idea of the economic and human cost of plant pests and diseases. This is precisely the area where plant genetic engineering seems to promise some dramatic improvements.

Viral diseases take their toll of plants, just as they do of humans. We have few defences against AIDS and the common cold, and plants have no self-produced protection against viruses. However, plants can take advantage of a phenomenon called induced resistance. The presence of proteins from the offending virus in the plant cell seems to ward off attack by full virus particles. No-one really understands how this works, although superficially it is like the process of immunisation that protects humans from viral diseases such as smallpox. Plants can be protected by injection with milder forms of the viruses that attack them. The genetic engineering approach is to transfer proteins from the coat of the relevant virus to plant cells in culture. The resulting transgenic plants are virus resistant. So far this has been shown to work for tomato and potato, both plants of immense commercial importance.

Transgenic tomatoes resistant to fungi have already been created, by borrowing genes for antifungal proteins that occur naturally in other plants, such as tobacco. The cell walls of fungi contain chitin, the second most abundant polymer on Earth (after cellulose, to which it is chemically related). The antifungal proteins include enzymes, called chitinases, that degrade the chitin and effectively destroy the cell walls of invading fungi.

Perhaps the most effective engineering strategy against plant invaders has been the creation of a whole range of plants with their own insecticide. It has been known since the turn of this century that strains of the bacterium *Bacillus thuringiensis* (bt) produce spores that contain protein crystals toxic to insects. The spores have been developed as an insecticide for protection of plants against caterpillars of moths and butterflies, mosquitoes and blackfly. Different bt strains produce toxins specific for different pests. Recently bt toxins against worms, beetles and mites have also been discovered.

The bt toxin spores have been one of the most successful weapons in the armoury of biological control, in which natural products and predators are used to fight plant disease – in place of synthetic chemicals. In parts of West Africa where bt has been used, river blindness, which is carried by blackfly, has been completely eliminated. However, bt still only accounts for about one per cent of the pesticide market, and is more expensive than chemical pesticides, although it has the bonus of being harmless to humans and other animals. Like other pesticides, the bt toxin has to be sprayed repeatedly, which pushes up costs.

The bacterial gene coding for bt toxin was first transferred to tobacco plant cells, which were regenerated into transgenic plants that expressed the toxin. These remained immune to an insect attack that devastated normal plants within a few days. Since these first experiments, bt toxins have been transferred to tomato, potato and cotton – the latter being an important cash crop normally subject to huge losses from insect attack. Despite the success of genetic engineering experiments with the bt toxin, it will still be years before the plants reach the market. Even then, it is hard to predict how well they will compete with chemical pesticides. There is also the problem that insects may become resistant to the bt toxin. The message is probably that a range of weapons should be used to help plants fight disease and predators, with biotechnology and genetic engineering adding to the available options.

Bt-containing plants have received a great deal of attention in the scientific community, but there are other ways of engineering plants to resist the attentions of insects. For example, US and Spanish biotechnologists have recently reported a way of protecting stored seeds from attack by weevils. Sometimes insects devour 100 per cent of seed stores, but they do less well if their diet contains an enzyme that stops them breaking down the starch in the seeds. Beans have the gene for this enzyme; peas do not. Transferring this useful gene from beans to peas gave transgenic pea plants that were safe from insect attack.

Weeds, competing for space, water and nutrients, can also be major enemies of crop plants. Existing herbicides are usually not specific enough to leave crop plants unharmed when they are applied to a plot. Hand weeding is labour intensive, and although there are biological toxins available, some companies are beginning to develop plants that are resistant to the application of herbicides. The most developed is a plant resistant to the herbicide glyphosphate. Known commercially as Roundup, this is the weedkiller you would use to clear a neglected garden or allotment. It works by blocking an enzyme that the weed needs to make amino acids. If some weedkiller splashes onto other plants, however, their enzymes will also suffer. The herbicide-resistant plants contain a gene for a slightly different version of this enzyme, which glyphosphate cannot touch. If a field of these transgenic crop plants is sprayed with Roundup, the weeds succumb but the crop plants remain unscathed. So far Monsanto, the company which makes Roundup, has engineered Roundup resistance into the usual model plants (tobacco, potato, and tomato) and has its sights set on major crop plants such as soya. Inevitably, Monsanto has had to counter charges that its motivation is to make farmers buy and use more Roundup. In reply, the company says that Roundup, which has low toxicity and is easily degraded, is kinder to the environment than other herbicides. Using Roundup, at the expense of alternatives on the market, should therefore have a beneficial effect on the surrounding environment and water supply.

Another strategy to boost the yield of crops is to use genetic manipulation to override the ecological limits of various plants. Imagine growing mangoes in Scotland or salads in the desert! This will mean taking a close look at the genetic basis of plants' tolerance of cold, drought, wind and various soil conditions such as salinity and acidity. Exciting progress is being made in understanding how plants cope at low temperatures. Avoiding cold and frost damage by planting the right species at the right time is one of the basic rules of gardening. You would never sow tomatoes in the autumn outside, and expect them to survive the winter, yet broad beans usually cope quite well under these conditions. It is the same story with frozen food: blackcurrants and peas freeze well, but no-one has found a really successful way of freezing strawberries or lettuce.

An important factor in resisting cold damage is a plant's ability to alter the composition of the molecules in its cell membrane. This is the barrier that separates the cell from its outer environment (in a plant cell it lies inside the cell wall). It contains a large number of oily molecules called lipids arranged in a double layer in which various protein molecules are embedded. This picture of a cell membrane is called the 'fluid mosaic' model. This is a good description of the way a cell membrane functions. It is a dynamic rather than a passive or rigid barrier. It lets substances in and out through channels in the membrane protein. The lipid molecules can move sideways, which gives the cell some much needed flexibility. For example, white cells in the immune system, called phagocytes, engulf bacteria with their cell membranes (prior to digesting them within the cell). They can do this only if the cell membrane has some mobility.

But these vital properties can be lost when temperatures fall. All molecules become less mobile as the surrounding temperature decreases. In fact there is an intimate connection between molecular motion and temperature. The laws of physics say that the lowest temperature achievable occurs when all molecular motion ceases. This temperature, known as absolute zero, is at −273 °C. It turns out that it is impossible to make molecules stay completely still; the best that can be done is to put them into a state where they have a tiny residual energy. For years scientists in the strange world of low temperature physics have been trying to get as near to absolute zero as they can. So far their best efforts hover within a few billionths of a degree of the magic −273 °C. To appreciate just how cold this is, note that the lowest climatic temperature on earth ever recorded was −89.2 °C, at Vostok research station near the South Pole in the Antarctic.

At any given temperature, the amount of mobility lipid molecules have depends upon their chemical make-up. For example, at room temperature in your kitchen olive oil is liquid, whereas butter is a soft solid. But when you take butter from the fridge it is hard, and difficult to spread. Lipids such as olive oil are said to be unsaturated, while the harder lipids such as butter are called saturated (the terms come from the way the carbon atoms in the molecule are joined to one another and hence how many hydrogen atoms can bond to the carbon atoms).

There is good evidence that some plants that survive low temperatures do so by shifting the balance of their membrane lipid composition towards the more fluid unsaturated lipids. Genes coding for an enzyme that turns a saturated lipid into an unsaturated one have been identified in many plants. What the cold-resistant plants seem to do is switch this gene on quickly as temperatures fall.

Some bacteria have similar genes for coping with the cold. In 1990, Japanese scientists transferred such a gene from one bacterial species to another that was originally temperature sensitive and had no unsaturated lipids in its membranes. Once it acquired its new gene its membranes started to accumulate unsaturated lipids, and it was able to function well at low temperatures. One day it might be possible to do this with plants and so extend the temperature range of valuable species downwards.

However well membrane lipids perform, another danger still threatens when temperatures approach freezing, and that is ice. Once ice crystals form in a cell it is doomed. Without liquid water, essential biochemical reactions cannot occur, and the ice itself damages the delicate inner structure of the cell. Some cells cope by manufacturing antifreeze molecules, which stop the ice from forming.

Genetic engineering has been used to help to protect strawberries from ice damage. The strawberry plant is vulnerable to frost damage because a bacterium called *Psuedomonas syringae* lives on its surface. The bacterium produces a protein that actually encourages ice crystals to form. Steven Lindow and his team at the University of California at Berkeley have engineered a strain of *P. syringae* with the gene for the ice-forming protein cut out. If they spray the strawberries with these so-called 'ice-minus' bacteria, then the strawberries are protected from the frost.

Resistance to poor soil conditions can be developed by selection of suitable plant cells in tissue culture. Although the usual aim of tissue culture is to obtain many plants that are genetically identical, in practice there is a tendency, known as somaclonal variation, for the cells to alter their genetic constitution while in culture. If a stress, such as high salt concentration, is applied at this stage, then survivor cells can be selected out and grown on into salt-resistant transgenic plants. This is the sort of work that has been carried out for many years by the Tissue Culture for Crops Project (TCCP) at Colorado State University in the United States. According to the TCCP, acid-resisting sorghum and salt-resistant rice are now undergoing extensive field trials.

Another strategy for developing plants with high resistance to environmental stresses is to combine resistant species with species that are high yielding. Under normal plant breeding conditions, you cannot breed from two distinct species. However, it is possible to fuse protoplasts from two species by treating them with various chemicals. A cell with the genes of both 'parents' is obtained. The plant obtained from this has usually lost a few genes as it develops from this cell, but with luck it will still retain the gene or genes of interest. Then conventional breeding of this plant with the desired crop plant will lead to a product with the desired characteristic. One success by this route has been the development of a salt-tolerant rice by fusing protoplasts from a wild rice grown in the salty mangrove swamps of Bangladesh with protoplasts of a food rice.

Genetic engineering could be used to add to the nutritional quality of existing crops. Seeds, which are the foodstore of the developing plant, are the mainstay of the human diet. Animal proteins, however, are superior to the plant proteins found in seeds because they contain all 20 of the essential amino acids that are the building blocks of proteins. Once ingested, the proteins are broken down to provide the ingredients for human proteins. Plant proteins are deficient in some of the amino acids. To prevent dietary deficiencies ethnic cuisines that include little meat will combine plant proteins from different sources – Mexican rice and beans, or Middle Eastern hummus and pitta bread, for example. Health conscious vegetarians and vegans do likewise with a peanut butter sandwich or beans on toast.

It may be possible to create plants that contain complete proteins with the full range of amino acids. Already the genes for plant proteins such as phaseolin, which is found in French beans, and zein, from maize, have been cloned. Maize proteins lack the amino acids lysine and tryptophan, while legumes are low in the sulphur-containing amino acids cysteine and methionine. It might be possible to combine both proteins in one transgenic plant – but this kind of work is still at a very early stage.

The creation of plants with distinctive health benefits is also on the agenda. One obvious way forward is to alter the composition of the oils in the seeds. Most health authorities now advise consumption of less fat in the diet to cut the risk of heart disease – the major killer in the West – and possibly of some forms of cancer. They also suggest a shift in the balance of the fats and oils that are consumed, away from saturated hard fats, such as those found in dairy products, some meats, and chocolate, towards the polyunsaturated fats found in vegetable oils (this is similar to the shift that helps plants to maintain the integrity of their cell membranes in the face of cold stress, as just discussed, but there is no obvious connection between the two phenomena).

The new healthy vegetable oils will be created in a similar way to the cold-resistant plants by transferring the gene that can turn saturated lipids into unsaturated ones. Preliminary experiments show that this approach works.

Sometimes genes are removed from a plant, rather than being added, to create a novel product. Flavr Savr, a transgenic tomato that is the first genetically engineered food to find its way into the market, has been modified in this way. Most fruits such as tomatoes are picked when green and made to ripen artificially using the gaseous hormone ethylene. Some horticulturalists say that this does not allow time for sugars and other flavour compounds in the fruit to ripen, and may be why consumers put tasteless tomatoes on the top of their list of complaints when surveyed.

However, if we did let fruits ripen naturally before shipping there would be more mouldy melons and rotten tomatoes because, after ripening, softening sets in, and this attracts bacteria and fungi. Flavr Savr, created by the American company Calgene, has had one of the genes responsible for softening turned right down.

This is done by using antisense technology – which was discussed in Chapter 2. The tomato cells are treated with a single strand of DNA for the softening enzyme, with the sequence reading 'back to front'. This 'antisense' DNA binds to the mRNA of the gene, so that it cannot be translated into the protein (which in this case is the softening enzyme). The tomato manages to produce only a meagre one per cent of the normal amount of the softening enzyme. So Flavr Savr may safely be left to ripen on the vine. The new tomatoes are more likely to be processed into purée and ketchups than to be sold for salads.

Blue roses and brick red petunias

One day genetic engineering may even find its way into the florists and flower markets. The world cut-flower and ornamental plant industry is growing in volume, with roses, carnations and chrysanthemums accounting for nearly half of all exports. People in the industry say that novelty is a major driving force in the world of flowers. So-called molecular flower breeding brings genetic engineering to the search for eye-catching new species.

The reason why you cannot buy blue roses or yellow cornflowers is that no one plant species has the genes for the whole spectrum of colour pigments. So conventional plant breeding is limited in the colours of flowers that can be produced. The way in which a plant builds up one set of pigments is fairly well understood. This set is called the anthocyanidins, which account for many of the red, blue and purple colours of both flowers and fruits.

The anthocyanidins are built up from the amino acid phenyl-alanine. Over the last 50 years or so, chemists from around the world have patiently worked out the steps and the enzymes that plants use to make each of the pigments. First the simple phenylalanine molecule is elaborated by a set of four chemical reactions to give a yellow pigment called a chalcone. The enzyme that makes the chalcone from the preceding molecule is called chalcone synthase (CHS) and it turns out to be rather important where flower genetic engineering is concerned.

What happens next depends upon which genes the plant has. There are three sets of genes that code for three different sets of enzymes. These act on the chalcone to make purple, blue or red pigments. Petunias do not possess one of these sets, and can never naturally be the brick red colour of geraniums, which do have these genes.

The situation is made even more complicated by the fact that flower colour can change depending on the acidity or alkalinity of the plant's sap, or even the soil in which they grow. Poppies and cornflowers both contain the 'red' pigment, but because the sap of cornflowers is alkaline the pigment turns blue. This ability of anthocyanidins to change colour is easy to demonstrate in a kitchen table experiment. Grind a few petals from a red rose, cornflower or delphinium with a small amount of water to make a coloured solution. Add a few drops of this to vinegar (acid) or baking powder (alkaline) and watch the colours change. You can get the same effect with ground up strawberries, blackcurrants and beetroot. Hydrangeas and bluebells change colour depending on the nature of the soil they are grown in. Ants produce formic acid, making bluebells grown on an ant heap pink, while the colour of hydrangeas varies from one garden to the next.

The first successful experiment to genetically engineer flower colour was carried out by scientists at the Max Planck Institute in Cologne. They transferred a gene for an enzyme that can make pelargonidin – the brick red colour of geranium – to petunias. Then a Dutch team, led by Alexander van der Krol intervened, at an earlier stage of pigment synthesis. They attempted to turn off the CHS gene using the kind of antisense technology used for making the Flavr Savr tomato. This puts a block on pigment production and the result was petunias with flowers that were much paler than their usual brilliant shades. Some were even pure white. An unexpected result of these experiments was petunias with unusually patterned petals – presumably because the level of CHS expression was different in different parts of the petals.

Recently the attention of the molecular flower breeders has turned to commercial varieties of chrysanthemums and roses. A popular pink chrysanthemum has been turned white – using experiments similar to those used to make the pale petunias.

Gene plunder and patents

There are over 300000 plants that have been identified and no-one is sure what percentage of the total this is. Areas of rich diversity, such as the tropical rainforests, almost certainly contain thousands of as-yet undiscovered plants. Mere identification of a plant does not mean that it has been fully investigated for its food value, products, or genes. Thousands of species are being lost as land is turned over to industrial and transport use, whether for building motorways in Britain or for developing cattle farms in Brazilian rainforests.

In one sense genetic engineers appear to be colluding with this wholesale destruction of plant biodiversity. The idea of searching out genes that can then be transferred to an existing crop plant reduces the source of the gene to just a piece of DNA. The implication is that once the gene has been cloned, and its sequence logged on to a database for everyone to look at, then its original plant home is of no further interest.

Nevertheless, the driving force behind plant biotechnology is the search for interesting genes. It is therefore in the interests of biotechnology that genes should be preserved, and this necessarily implies the need for worldwide agreement on biodiversity.

The increasing emphasis on genes has led to fears in developing countries that Western scientists could rob them of their most precious commodity – their plant genes. Countries such as Ethiopia, which is particularly rich in coffee and cereal species, have led the fight against gene plunder by forbidding export of germplasm. India has just passed a tough new law that will stop drug companies raiding the country for medicinal plants and microbes from which to extract profit-making medicines.

This conflict has been heightened by the prospect of patenting genes and their products such as transgenic plants. Plant varieties obtained from traditional breeding were never granted patents in the way other inventions are. Instead they were protected by something called plant variety rights, which tried to strike a balance between repaying the original breeders for their investment, and relieving the royalty burden on those who wanted to do further breeding with these varieties (so-called 'breeders' privileges'). Similarly farmers were allowed to save some seed from crops from a protected variety without paying further royalties.

Because transgenic plants are more truly novel than varieties obtained via traditional breeding, the authorities in Europe decided to grant them full patent protection. But no-one is yet sure – because we have so little experience of transgenic plants in the marketplace – just how this will affect breeders' and farmers' privileges.

It is easy to sympathise with people in the countries from where genes that end up in transgenic plants originate. The biotechnologists, having obtained patents on the plant, will extract royalties from their customers, with no obligation to give anything back to the place where they found the valuable gene. They could, of course, argue that without substantial investment, the gene could not have been transformed into a money-spinning product.

The International Undertaking for Plant Genetic Resources, which operates under the direction of the United Nations Food and Agriculture Organization has argued that genes are the common property of humanity. There are moves to draw up some equitable agreements for sharing the profits from transgenic plants between the country where the genes originated, and the country developing the plants. Many international companies are now anxious not to be labelled as 'gene thieves'. Up until a few years ago, it was common practice for employees of pharmaceutical firms to bring a handful of soil or a few cuttings home from holidays abroad to screen for interesting drugs. This has stopped, and explorations of this kind are carried out in collaboration with scientists in the country concerned.

More food for thought

The manufacturers of genetically engineered food know they will have a tough job convincing the public to buy their products. Consumers in the United States and in Europe are demanding more and more information about what is in their food. Many of them reject highly processed foods, and 'natural' foodstuffs such as organically grown vegetables are gaining in popularity. Added to this are strong pressure groups who promote the view that genetically engineered food is both unnatural and dangerous.

Inevitably this has led to consumer pressure that genetically modified products such as the Flavr Savr tomato should be labelled so that informed choices can be made. In the USA, foods are regulated only according to their safety and quality. In Europe there is the concept of a novel food, which may relate to the way it has been produced, even if the end product is the same as that made in the conventional way. A detailed report on the Flavr Savr tomato for the United States Food and Drug Administration showed no significant differences in content between it and a conventional tomato – except for one thing. Flavr Savr still has kanamycin resistance genes, which are used to select out the genetically engineered plant cells at the start of the modification process. Nutritionally these have no effect, but some opponents argue they could end up inside the bodies of consumers and make them resistant to kanamycin (an antibiotic that is, in fact, hardly ever used because of severe side-effects). They would then be, in theory, vulnerable to infection. It is overwhelmingly likely that the kanamycin resistance genes do not survive the acid in the gut. Nevertheless, maybe for some consumers, the vanishingly small risk that they might do is just unacceptable and they feel they have the right to know that these genes are present in tomatoes they might purchase.

One thing consumers should bear in mind, however, is that antibiotic resistance genes are not confined to organisms that participate in genetic engineering – they are widespread in the bacteria that colonise so-called 'natural' foods. This is because antibiotic resistance is a natural property of microbes, as discussed in Chapter 4. Of course there is bound to be DNA from other organisms in every plant or animal food we eat – because it is present in all cells.

We can be sure, however, that there will be antibiotic genes in genetically engineered products. So it is clear that the process by which it is produced can affect the content of a food. The argument around labelling centres on whether the labels should relate just the nutritional content of a food, or whether it should also give information about how it was produced. This is an argument that will not be resolved easily. There is a good argument for keeping labels as simple as possible – just look at the labels on so-called 'cruelty free' cosmetics and toiletries to see how cynical and misleading the labelling game can be. Yet there is an equally good case, on ethical grounds, for giving consumers the maximum possible information, so they can make a free choice. The most obvious aspect of this debate is that the consumer benefits most from whatever labelling system is decided on if he or she has a clear understanding of the science involved in making the product offered.

But even if genetically engineered food turns out to be perfectly safe for the consumer there are bound to be wider concerns about the effect of transgenic plants on the environment. Probably the main worry is that if we make crop plants fitter, by adding pest and disease resistance genes, they will transfer these traits to weeds, which will then take over the crop plants' ecosystem.

Weeds usually do well because they have a particularly beneficial combination of genes. Adding genes that represent a drain on the weed's biochemical energy will not help its career as a weed. A gene from a transgenic plant is particularly unlikely to be received as a welcome present if it confers resistance to a pest to which the weed is not susceptible.

However, it must be said that we have very little experience of how transgenic plants behave in the field. There are many small- and medium-scale field trials going on, with no reported adverse effects on the surrounding ecosystem. Perhaps the most interesting feature of these trials is that in Britain at least they are open for public comment as from two years ago. Although details of forthcoming field trials of transgenic plants must be published in newspapers, few people have commented about the issues involved. This is a pity because, if the public was to work with the scientists on assessing the risks to the environment from transgenic plants, the benefits of plant biotechnology to society could be realised more easily, and the potential dangers neutralised at an early stage.

11

Environmental solutions

Concern about the state of our environment has slowly worked its way up the political agenda over the last decade or so. This new awareness exists on many levels: there are global issues such as climate change, local problems such as traffic congestion, as well as health worries caused by toxic waste and air pollution.

Biological processes, such as photosynthesis, helped to shape our environment in the early days of evolution, as we shall see in Chapter 12. The aim of environmental biotechnology is to use the transforming power of biochemistry – driven, of course, by DNA – to help to create an environment that we would be proud to leave to our grandchildren.

There are two main ways in which biotechnology can soften the impact of human activity on our planet. It can help in the provision of energy and material resources, and in the destruction of pollution.

Trapping energy from the Sun

As the world's population grows, so do people's aspirations. Hence many people in developing countries want the cars, fridges and other objects that we take for granted in the Western world. At the same time, they also want to develop their industries to generate income and a higher standard of living. Inevitably this will result in increased energy consumption – from more electricity to drive domestic and industrial appliances to more fuel for heating and transport.

Today the commercial energy market is dominated by coal, oil and gas – the fossil fuels. These are the ultimate products of photosynthesis, the process by which the energy of sunlight 'fixes' the carbon dioxide in the atmosphere in the form of more complex carbon-based molecules such as glucose, starch, fats and proteins in plants (see also Chapter 10).

When living things die and decay, assisted by microbes known as decomposers, the carbon may re-enter the atmosphere between organic (living) and inorganic (non-living) forms. The level of carbon dioxide in the atmosphere due to these biological processes is around 280 parts per million (p.p.m.) or 0.028 per cent. However, under certain geological conditions, the carbon-based remains of living things are transformed, over millions of years, into fossil fuels. Coal contains a high percentage of pure carbon, oil is a complex mixture of hydrocarbons (compounds containing carbon and hydrogen only) whereas natural gas is mainly methane – the simplest hydrocarbon.

When fossil fuels are burned, their chemical energy is released as heat and light. This can be used either directly – as in a gas cooker – or indirectly to drive electricity generators or vehicle engines. Since the Industrial Revolution, the burden of carbon dioxide in the atmosphere due to these so-called anthropogenic emissions has been added to the amount resulting from biological processes. The total level of carbon dioxide is now 350 p.p.m. and this will rise to around 560 p.p.m. within the next 40 to 50 years, if energy consumption grows at its expected level.

This is still a tiny proportion of our atmosphere, which is composed mainly of nitrogen and oxygen. However, even a small amount of carbon dioxide can have a profound effect, because it is a so-called greenhouse gas. Carbon dioxide has always played an important role in maintaining the Earth's temperature at a level comfortable for life. The tiny amount present in the atmosphere traps enough heat from the Sun to keep the average temperature of our planet at 15 °C. The higher projected carbon dioxide levels mentioned above will probably raise this temperature by one or two degrees. It does not sound like much, but it might have a profound effect on living things – from increasing the incidence of water-borne disease and flooding to boosting the yields of crops such as tomatoes and the range of butterflies surviving the winter in the inner city. This is global warming. Unlike the greenhouse effect, which is the natural trapping of heat by carbon dioxide, it results entirely from human activity.

When fossil fuels are burned they are lost – at least in the short term. The carbon dioxide they release will eventually find its way into plant material through photosynthesis and this will in turn decay into fossil fuels. In the meantime we have global warming, and the problem of dwindling supplies of fossil fuels.

A small, but increasing, proportion of electrical energy is supplied from renewable resources such as the wind, waves, and falling water (hydroelectric power). Unfortunately these are not as energy dense as fossil fuels. You need a few thousand wind-powered generators to get as much energy as one coal-fired power station, for example. But these renewables do not produce carbon dioxide.

Nuclear power is energy dense, and does not produce carbon dioxide – but it has other drawbacks. It excites more public opposition than any other source of energy because of fears about the dangers of the radiation it produces. There is no easy solution to the problem of storage and disposal of nuclear waste.

So, as no energy source is ideal, it is important to develop a range of options. Biomass is plant material grown for the production of energy. In developing countries, outside the commercial energy market, it accounts for much of the local and domestic energy consumption – in wood-burning stoves for cooking meals for example.

Wood is the biggest, and most economic, source of biomass, but trees may take years to reach their full height. Interest in developing biomass as a credible and commercial energy source has focussed on selecting fast-growing species that burn efficiently. For instance, *Leucaene leucocephala*, native to Mexico, is now being grown in India and is already supplying local people with energy only three years after being planted. There are many other species with similar potential, for example, *Eucalyptus* and shrubs such as *Euphorbia*.

Direct burning is not the only way to extract energy from plants. For thousands of years yeasts have been dismantling plant sugars and starches to obtain the biochemical energy they need to live, and producing ethanol (one of a family of chemicals called alcohols) as a waste product. For humans, ethanol is a social lubricant, and a legal drug. Until recently, the only evidence of its fuel value was the fierce and ephemeral flame of the brandy burning on Christmas pudding or Crêpe Suzette.

Only minor adaptations to engines are needed for cars to run on a mixture of ethanol and petrol. Brazil took the lead in this area, producing ethanol from sugar cane to sell as a petrol additive in 1975. New cars can be made with engines designed to run on pure ethanol. Sugary and starchy crops, such as sugar beet, potatoes, cassava and Jerusalem artichokes, have the best potential for ethanol production. Making it from cellulose-based materials is more demanding, because of the need to convert the cellulose first to starch, prior to fermentation.

Petroleum, or crude oil, supplies fuels for a wide range of uses – for cars, ships and aeroplanes as well as central heating systems. The challenge is to see whether plants can match this. Already it seems that they may be able to. Besides direct burning and ethanol, there are also plant-based 'diesel' fuels. These are plant oils such as sunflower, rapeseed and olive oil. There is even a petroleum nut in the Philippines that burns brightly when it is ignited, and botanical research suggests that there might be over 300 different plant species that can give diesel substitutes.

Rudolf Diesel, the inventor of the diesel engine, used peanut oil to fuel one of his engines at the Paris Exhibition of 1910, and in 1912 he wrote, 'The use of vegetable oils for engine fuels may seem insignificant today. But such oils may become in the course of time as important as petroleum and the coal tar products of the present time.' Buses, taxis and lorries in towns could run on biodiesel now; the only barrier is financial. In Europe, Austria has led the way in biodiesel. Elsewhere plant-based fuels still need the sorts of tax incentive that were given to unleaded petrol in the UK a few years ago (which led to a dramatic fall in lead levels in urban air).

You do not even need to grow plants specially for fuel production. For many uses, waste material does the job equally well. Molasses from sugar production, rice husks, cheese waste and citrus fruit skins have all been used for ethanol or direct energy production. Agricultural waste such as cattle dung and straw can also be burned to produce cheap energy and there are schemes based on these ideas up and running all over Europe.

Rubbish tips are an eyesore. They are also cosy niches for communities of bacteria and fungi. In the low oxygen environment of a landfill site, species known as methanogens flourish. As their name suggests, they produce methane as they feed off the organic material in the rubbish, often after it has been part-processed by other microbes (see also the discussion on archaebacteria in Chapter 4). Landfill gas is a mixture of roughly equal volumes of methane and carbon dioxide that is produced by this microbial activity. One tonne of waste produces 100 times its own volume of gas over a ten year period. Inevitably, as methane is flammable, this is a hazard. In 1986, a bungalow in Loscoe, Derbyshire, was demolished – injuring the occupants – by a natural explosion at a nearby landfill site.

For safety and economic reasons, engineers began sinking pipes, wells and boreholes into landfill sites to tap off this gas, which can be used for local heating or even electricity generation. It makes sound environmental sense to eliminate the methane by burning it to give carbon dioxide. Methane, like carbon dioxide, is a 'greenhouse' gas, and contributes to global warming, but, per molecule, it has 27 times the warming potential of carbon dioxide. There is a great deal of unexplored potential in landfill. Microbiologists reckon only a third of the possible gas is being generated. Better arrangement of the rubbish in the site, and more research into the types of microbe working there, could really boost the prospects of this type of energy generation. Of course, it is always possible merely to burn the rubbish, and use the heat to generate power – as is done at many commercial and public incinerators.

Probably the best that biomass can do is to diversify energy supply, and provide cheap low-tech energy for local firms, communities and projects. All direct burning of organic material will produce carbon dioxide, so in this sense biomass poses a similar threat of global warming in the short term as fossil fuels (although plants do absorb carbon dioxide and so help to balance the equation). In fact the danger of atmospheric pollution can be greater if rubbish is burned. The acrid fumes of a bonfire contain a cocktail of waste gases, some of them highly toxic. But other biofuels, such as plant diesel and landfill gas, have the advantage of not producing sulphur dioxide, a major pollutant that comes from burning coal.

Microbial mining

Besides energy, one of the other major requirements of human civilisation is a supply of metals. Well over half of the 92 naturally occurring elements are metals. Many of these combine toughness and strength with an ability to be shaped into objects as diverse as a car, a crane, or an artificial limb. Although metals have been replaced by plastics for thousands of uses, doing without metals is as unthinkable as getting by without electricity.

Most metals are reactive elements. If, for instance, a tiny lump of the metal caesium is dropped into a tank of water the resulting explosion is one of the most spectacular in chemistry. Needless to say there is no market for cars made of caesium, but the metal is used in the photocells of light meters. Less dramatic – and economically more serious – is the slow corrosion of iron to rust when it is exposed to air and water.

The consequences of the chemical reactivity of the metals is that since the Earth was formed around five billion years ago, they have combined with other elements – mainly oxygen and sulphur – and embedded themselves in the Earth's crust as minerals. To extract the pure metal involves mining the mineral and then, in most cases, chemical or electrical processing.

Mining and mineral processing have always had an environmental price tag. First, mines leave an indelible mark on the landscape. Second, extractions such as the production of aluminium from bauxite use up large amounts of energy and material resources. Polluting by-products such as sulphur dioxide might also be generated by metal processing. Even the metals themselves may contaminate the soil and water around the site of production.

There is an increasing urge to dig deeper and wider to extract metals – despite any attempt to recycle what we have by collecting aluminium cans and the like – driven by increasing population and industrialisation. The world's last great wilderness, Antarctica, is under great pressure to open up its mineral deposits. The continent is safe for the moment, thanks to a 50 year moratorium on mining imposed by the Antarctic Treaty in 1991, but demands from developing countries for more metals could place this agreement under threat. For instance, suppose all the Chinese were to demand the small amount of tungsten required for a bicycle lamp (a legal requirement in the West)?

To avoid plundering Antarctica – and carving up other parts of the planet with the messiness of mining – it might be in our interests to look for technology that can make the most of low grade minerals. Microbes offer one way of doing this.

The Romans extracted copper from the Rio Tinto mine in south-west Spain over 2000 years ago. The site is surrounded by pools of liquid, whose dark blue colour is the tell-tale sign of copper sulphate. Bacterial action leaches copper from the rocks of the mine into these pools. Research has shown that the microbe responsible is *Thiobacillus ferrooxidans*. This is a bacterium with a strange diet. Instead of obtaining energy by breaking down glucose – or other carbon-based food – it consumes minerals containing iron and sulphur, rather like a sideshow freak who swallows safety pins. These minerals are converted into other chemical forms by the enzymes of *T. ferrooxidans* and energy is released as part of the process.

The sulphur atoms in the minerals are converted to sulphuric acid, a powerful chemical that releases the copper present in the rocks in the form of copper sulphate and this dissolves in any surrounding water. No human labour is involved – *T. ferrooxidans* does all the work.

If you dip a coin into a copper sulphate solution, it readily acquires a coating of freshly deposited copper. A similar process is used to extract copper metal on an industrial scale from the Rio Tinto and other copper mines. So successful is the leaching of copper by *T. ferrooxidans* that the process accounts for one third of all copper mining worldwide.

The diversity of microbes is so rich that it would not be at all surprising to discover other species with the metal-extracting ability of *T. ferrooxidans*. Indeed similar systems are under development for the extraction of gold and phosphates (the latter are used in the manufacture of detergents, fertilisers, and soft drinks such as Coca-Cola).

Microbes also play a role in the recycling of metals. A certain species, whose name is currently a commercial secret, has been found to consume the paint layer sprayed onto the outside of aluminium cans in a matter of minutes. Not only does this speed up the recycling process, it sidesteps the need for solvents – which might be toxic – to dissolve the paint.

Alternative oil fields

Petroleum is the driving force behind both industrial and developing economies. Roughly half is used for energy and, as discussed above, at least some of this is now being replaced by plant-based fuels. The other half is used as a feedstock for the petrochemicals industry and here, too, plant materials are showing their enormous potential as a replacement for petroleum. Although the exact proportion varies from place to place, roughly half of petroleum after refinement is used for vehicle fuel.

The petrochemicals revolution has its roots in the 1940s and there can be few places in the world that have not been touched by it. It has been the basis of the modern pharmaceutical industry, replaced paper and wood with plastic, soap with detergents, and natural fibres with synthetics – to name but a few of its applications.

Plant oils contain compounds containing carbon, hydrogen and oxygen, known as fatty acids, which can stand in for the hydrocarbons of petroleum in many of their industrial roles. For instance, palm oils can make detergents, jojoba oil is a lubricant, and ricinoleic acid, from castor oil, has hundreds of industrial uses including the synthesis of cosmetics, plastics and nylon. The potential market for ricinoleic acid in the UK alone is £50 million.

What has stood in the way of the industrial development of these plant oils is the difficulty of cultivating the plants themselves. For instance, the castor oil plant is no more than a minor crop because it gives such poor yields, and the seeds of the plant contain not just castor oil, with its precious ricinoleic acid, but ricin, one of the world's most potent toxins (it was used to murder Bulgarian diplomat Georgi Markov in London in 1978).

However, there is good evidence that ricinoleic acid is related to one of the main components of olive oil and sunflower oil, oleic acid, by just one chemical step that can be carried out by the action of a single enzyme. Introduction of the gene for this enzyme into high yielding sunflowers could produce abundant amounts of ricinoleic acid for industrial use. British scientists are already working along similar lines. They are trying to transfer an enzyme to oilseed rape that would convert its oleic acid into the closely related petroselenic acid. This acid, found in coriander seeds, can be processed into a wide range of plastics and polymers.

Plants are natural factories for oils, and many other products. They can also be made to manufacture substances that originate in other organisms. One of the first examples of this was a biodegradable polymer, polyhydroxybutyrate (PHB). This is a natural product of a bacterium, *Alcaligenes eutrophus*, but has proved difficult to process after fermentation. Chris Somerville of Michigan State University has shown that PHB, which is already used in Europe in objects such as shampoo bottles, can be produced in *Arabidopsis*. Other researchers are aiming to produce it in potatoes in place of starch granules.

High yielding sunflowers could also be a home to genes for natural materials that can replace petroleum-derived products. The silk that is spun by spiders is as strong as synthetic fibres, and far more elastic, but you would need 400 spiders to get a square metre of cloth woven with the silk. The genes that make the silk, however, have been identified. So they could be transferred to the sunflowers, making them into 'factories' that would spin the spiders' silk.

You can even use plants as bioreactors for the production of therapeutic proteins. Examples include interferon, used in cancer treatment, and α-amylase, which is used in baking as well as in human and animal vaccines. An obvious advantage of the plant bioreactors is that large amounts of products can be manufactured, and, as with animal bioreactors, the correct modifications such as glycosylation can be carried out by the plant cell's biochemical apparatus.

Some of the first discoveries in this field came as a bonus from the search for new ways of enhancing a plant's resistance to disease. For instance, when researchers at the Scripps Research

Institute in California tried to make a plant produce antibodies against attacking pathogens they realised that mouse antibodies could easily be produced by the same plants.

Like the industrial oils discussed above, products from plant bioreactors give farmers new options in an era when food over-production is widespread and represent a new departure in land use.

The biotechnology clean-up squad

It is impossible to keep soil, water and air in pristine condition in a modern industrial society. Even tightening environmental standards cannot protect us from the legacy of a dirtier past, and pollutants know no national boundaries. The success of legislation in favour of unleaded petrol was measured by looking at the fall in lead levels in the Arctic, where readings are unlikely to be confused by background environmental signals. The lead had migrated to the North Pole, over the years, from the car-owning nations.

People worry incessantly about pollution, and want guidelines and priorities. Pollutants are chemicals that are perceived to have some adverse effect on human health or on wildlife. Because humans have such complicated lifestyles, it is difficult to link ill health and increased mortality to specific pollutants. Often evidence for the harmful effects of chemicals is gathered from the medical records of industrial workers who have been exposed to a far higher level of a chemical than the general public ever would be, under normal circumstances.

For instance, the higher than average frequency of bladder cancer among people working in the rubber industry was a warning of the cancer-causing potential of the hydrocarbon benzene. This, along with other hydrocarbons, is present in diesel fumes. But there is no hard evidence that people who spend much time cycling behind buses are more likely to get cancer from their exposure to benzene than people who are not exposed to diesel fumes.

This does not mean that benzene is innocent as far as the

average person's health is concerned – just that evidence of guilt is hard to come by. Until recent years there was no need to worry about the potential harm done by low levels of benzene, or other chemicals, in the environment, because they could not be detected. Now sensitive monitoring equipment can detect femtogram levels (that is 10^{-15} or one thousand million millionth of a gram) of such substances in samples of air, water or soil. Biotechnology has played its role in the development of environmental monitoring – enzymes, antibodies and PCR (see p. 150) are all used to detect various kinds of contamination such as *Salmonella* bacteria in food or *Legionella* (the microbe that causes Legionnaire's disease) in central heating systems. Governments can set safe limits, based on the available evidence, and these limits are revised upwards, or downwards, from time to time.

Nevertheless, it is hard to give definite advice on the risks of environmental pollution. Some people play safe and wage a relentless campaign against all 'chemicals' and 'germs' in the environment and the media too often plays on these fears, but there is too much uncertainty for industry to adopt a complacent or defensive attitude – although it does on occasion.

Inevitably, given the discussion above about their importance to society, metals are widespread environmental contaminants. It is the so-called heavy metals – lead, cadmium, nickel, and the radioactive elements such as uranium and plutonium – which give cause for concern. These are probably toxic because they stand in for vital metals, such as iron, that are used in biochemical processes.

However, heavy metals are not toxic to some microbes and plants. Many fungi can trap heavy metals on their cell walls, in a process known as biosorption. The walls are rich in a polymer called chitin and it is this molecule that binds tightly to the atoms of the heavy metal.

Hundreds of experiments have shown that biosorption works. Fungi can be pressed into a mat over which toxic waste is allowed to trickle. The heavy metal concentration is dramatically reduced as chitin picks up metal atoms. Chitin is also found in the shells of crustaceans, and other experiments have shown that prawn shells, discarded from the fish industry, can be equally effective. In fact, biosorption may be somewhat too successful – there have been warnings about eating shellfish from metal-polluted waters, because the shells may absorb the pollution and this could be ingested.

Biosorption does not make heavy metals disappear. It concentrates them and makes them easier to handle. After a typical biosorption treatment, you have a fungal filter loaded up with heavy metal. Treatment with chemicals such as sulphuric acid will release this metal for controlled disposal, or for recycling. The nuclear industry is investing in research into biosorption, and similar processes, because plutonium and uranium can be processed in this way. So biosorption adds a much needed technology to the difficult technical problem of how to handle and dispose of nuclear waste.

Plants too can handle heavy metal pollution. These plants, known as metal hyperaccumulators, cover the whole range – from herbaceous species to trees. One striking example is *Sebertis acuminata*, a native of New Caledonia where it flourishes on nickel-rich soils. The tree contains a latex, like rubber, which is blue because it contains more than 11 per cent nickel, which it has soaked up from the soil.

Experiments carried out by Professor Steve McGrath and his team at Rothamsted Research Station in the UK showed that alpine pennycress accumulated 100 times more zinc than oilseed rape. Another hyperaccumulator turned out to be the common rock plant alyssum.

Some microbes, and plants, specialise in dismantling carbon-based toxins. They turn the toxin into a non-toxic substance – usually carbon dioxide. These toxins are usually xenobiotics, chemicals that do not occur in nature, such as various solvents, plastics and pesticides that are manufactured from petroleum. Most governments have a list of the worst offenders, such as the polychlorinated biphenyls (PCBs), once widely used in the electrical industry and now decaying slowly in the environment to give a range of other toxic chemicals.

Some of these toxins are harmful to all living things, because they attack some of the basic functions of the cell, but chemicals that harm the nervous system probably would not harm bacteria, fungi or plants. For each chemical on the danger list, there is probably at least one microbe that will metabolise it – dismantling the sometimes complex molecules to reach the energy stored in its chemical bonds.

The trick to finding these handy microbes is to look at soil that is contaminated with the forbidden substance. Normally soil teems with microbial activity – there are around one billion bacteria to each gram. Contaminated soil is a specialised ecosystem that favours the survival of those microbes that can adapt to the presence of a pollutant. These species are isolated from the soil by standard microbiological techniques. Then, like the prospective transfer of the gene for ricinoleic acid synthesis to sunflowers, genetic engineering might be appropriate. The microbe with the gene for dealing with the pollutant might not be the best possible for survival in that soil. So the gene could be transferred to a 'better' species. Also, it is fairly unusual to find just one species carrying out the detoxification – usually a team of species work in relay. Recently, a group of microbiologists at the University of Minnesota managed to combine a set of 'detoxification' genes into one microbe. This, a *Pseudomonas* species, was able to remove polyhalogenated compounds, traditionally very hard to break down, from soil with high efficiency. They could even break down chlorofluorocarbons.

Plants too can destroy toxins such as PCBs. When people worry about the effects of toxic chemicals, they are probably not aware of the powerful protective mechanisms in their bodies. The immune system protects against pathogenic microbes, while the liver, and to some extent the kidneys, act on toxins. We all know that the liver sets to work on the alcohol we have consumed. It does a similar job with other chemicals from which the cells need protection. It does this by turning on the genes that synthesise a group of enzymes called P450s. It now seems that plants produce P450s as well, acting as a kind of 'green liver', which can dismantle forbidden chemicals such as PCBs and related compounds.

Biotechnology can even make existing efforts to help the environment more effective. Canadian and US scientists reported recently that enzymic treatment of recycled paper naturally sorted out the paper into two sorts of fibres – with and without ink.

Normally you would have to apply de-inking chemicals before pulping the paper for recycling. This creates the need for treating the waste stream from the recycling plant. With the cellulase treatment, no chemicals are used, and preliminary research suggests there could also be significant energy savings.

From agrochemicals to biotechnology

Agricultural chemicals – pesticides, herbicides and fertilisers – have brought enormous benefits, from boosting crop yields, to wiping out swarms of malaria-carrying mosquitoes. However there is little doubt that they have also harmed the environment. There are two main problems: they contaminate soil, water and food, sometimes persisting for many years; and they are not very specific, often wiping out innocent species as well as their targets.

It was Rachel Carson who first sounded the alarm with the bestseller *Silent Spring* in 1962. Since then, pressure groups and the general public have been campaigning against agrochemical pollution. Their efforts have led to bans on some of the most toxic pesticides, such as DDT, and controls on the levels of nitrates (which come from chemical fertilisers) in water.

Add to this the failure of pesticides to solve the problem they were created for. Fourteen per cent of crops are still lost to pests. Some pests – such as worms, which eat their way through a staggering $30 billion worth of crops per year – appear to have won their battle against the chemical industry, for many companies have now pulled out of this area.

Many scientists feel we should look at how pests are controlled in nature, and use biotechnology to develop these solutions. Before the intervention of the chemical industry, ecosystems waged war on their own terms, using a wide range of biochemical weapons. For instance, pyrethrins occur in chrysanthemums and are highly effective at paralysing and killing a range of insects. There are also microbes that attack insects, worms that kill slugs, and fungi that kill other microbes – all by manufacturing their own toxins.

There are three ways that biotechnology could harness these

natural abilities. The first is simply to redirect the attacking organisms to where they are most needed. For example, the UK firm Microbios makes sure that the nitrogen-fixing *Rhizobia* meet their leguminous plant partners by selling them to farmers. And there is new hope for anyone whose gardening effort has been devastated by slugs: Microbios is growing in fermenters a worm that attacks these voracious pests.

The second approach is to dispense with the attacking organism itself, and just extract the toxic chemical that it uses to annihilate its prey. Experiments by leading plant pathologist Gary Strobel of Montana State University have led to the isolation of the first plant-specific weedkiller. Weeds are more than an infuriating nuisance in the garden, they are a menace that leads to the loss of around a third of the world's crops. Chemical weedkillers can be effective, but they tend to attack other plants too. Strobel took an ecological approach to finding a natural killer for North America's worst weed, spotted knapweed. His dedicated colleague Andrea Stierle spent her honeymoon looking for sick spotted knapweed, and took diseased specimens back to the laboratory for analysis. It turned out that these plants were infected with a fungus that made a compound called maculosin, which was toxic to them. Strobel and his team set about extracting maculosin from the fungus, and hope to develop it as a biological weedkiller.

Finally, the most sophisticated way of developing natural pesticides and fertilisers involves genetic engineering. This means giving the organism under attack – usually a plant – the genes of a friendly species that could protect it, if only it were in the vicinity. Similarly, genes from a plant that can resist a particular threat could be transferred. The development of plants that contain the gene for an insecticidal protein (bt) has already been discussed in Chapter 10. There are many other defence genes being introduced to plants. Axis Genetics, in Cambridge, has identified many plant genes that code for protective proteins, and is busy transferring these to plants of commercial importance.

Even herbicide-resistant plants – often thought to have the potential to encourage the use of agrochemicals – may turn out to be 'green'. Nilgun Turner and co-workers at Rutgers University have recently created plants resistant to the herbicide bialaphos.

The plants they used were grasses such as the creeping bentgrass, which is often used on golf courses. Apparently bialaphos is preferable to other herbicides such as 2,4-D because it is broken down quickly. So if golf course weeds could be treated with this herbicide because the grass was resistant, this should avoid the build up of toxic herbicide.

So far, biotechnology-based methods have made little impact on the market for agricultural chemicals. The world spent nearly $49 billion on agrochemicals in 1994, and only $120 million on biopesticides and $50 million on biofertilisers. This is because biotechnological products, from transgenic plants to extracted compounds like maculosin, take a long time to test. But the balance between biological and chemical approaches to getting the best out of agriculture is bound to change over the next few years.

Biotechnology – is it really 'green'?

Few of the solutions that biotechnology offers for environmental problems have been put to the test on a wide scale. Genetic engineering – potentially of great value in applications such as cleaning up oil spills and contaminated land – has been applied hardly at all because of regulatory requirements that are driven partly by public fear of the release of genetically modified organisms.

However, there is a great deal of research and development going on. Most scientists do not expect biotechnological solutions to take over from conventional methods. 'Natural' pesticides and genetically engineered plants will, they say, account for only 20 per cent of the market at best.

However 'green' the image of environmental biotechnology, the current mood suggests that environmental activists and the scientists share little common ground. There is deep suspicion among the environmentalists about the 'biotech' way. In part, certainly, this is based on ignorance. Nevertheless, when one of their own begins to question biotechnology it must be time for the scientists to question their approach. A few years ago a prominent US plant molecular biologist, Martha Crouch, complained in a leading journal about the 'quick technical fix' that biotechnology offers to environmental problems. The biotech way just encourages people to think they can pollute the environment and a 'natural' solution will be found. Crouch would like scientists, politicians and the public to make some more radical choices about the environment. To this end, she left her post as a molecular biologist to dedicate herself to a less human-centred kind of science.

PART IV

The final frontier

12
Beyond DNA

The knowledge of the structure and function of DNA is probably the most powerful concept in biology, standing, as it does, at the very heart of our understanding of inheritance, and co-ordination of the biochemical activity of the cell. For some scientists, the molecular approach to the science of life – with DNA as the master molecule – provides a comprehensive understanding of nature. Unsolved problems such as the nature of human consciousness or how an embryo develops will be clarified as soon as the appropriate genes are cloned – they say. Others argue that molecular genetics is only part of a far larger picture and that other theories and ideas are equally worthy of attention and further exploration.

Neo-Darwinism and the selfish gene

Neo-Darwinism, as a theory, is not as modern as it sounds. The term was first coined in 1896, and refers to the synthesis of the work of Darwin with that of Mendel. Just to recap, Darwin's observations of nature led him to propose that variation of pheno-type within species occurred, and the variants that were best adapted to the environment would leave more offspring ('the survival of the fittest'). Thus, life evolved through the generations. Mendel took one step further towards a mechanism for inheritance with his experimental discovery of genes as being discrete inherit-able entities.

The discovery of the self-replication of DNA, and of gene mutations that were transmitted from cell to cell, and through the generations, provided a powerful explanation for both Darwin's and Mendel's theories at a fundamental molecular level. Genes are stretches of DNA. When they are expressed, through molecular mechanisms that are reasonably well mapped out, they give rise to a range of phenotypes. Mutation changes genotypes, which in turn alters the expressed phenotype.

The most outspoken advocates of neo-Darwinism today are Stephen Jay Gould of Harvard University, and Oxford scientist Richard Dawkins. Gould tends to defend the broad sweep of evolution – as in his analysis of the Burgess Shale and the explosion of life around the Pre-Cambrian period of 580 million years ago. There is plenty of defending to be done; you may think Darwinism is well established (after all it has been around for over 100 years) but there is no shortage of critics. These range from people who think the Earth is 5000 years old, to those who (quite rightly) point out the very real flaws in Darwin's original ideas.

Dawkins has developed a theory of his own, that of the 'selfish gene'. In his book of the same name he develops the rather bleak vision that humans – and other animals – are mere 'survival machines' that house the all-important genes. This is reminiscent of Tom Kirkwood's 'disposable soma' theory (see Chapter 8) of ageing. Dawkins uses examples from the world of animal behaviour to back up his theories.

Inevitably Dawkins' views have been labelled as reductionist, further reducing the dignity of humans! He, in turn, uses the 'selfish gene' theory as a platform to attack theology, arguing that it has no place in a world where life is driven by the blind urges of DNA to replicate. Curiously, however, Dawkins' ideas are almost religious in one sense – as some critics have pointed out. He places so much emphasis on DNA that the molecule almost has the status of some kind of 'vital' force. In other disciplines, like yoga or acupuncture, this might be called something else – like *prana* or *ki*. And the concept of some vital force flowing through biological material has a very long history, although it is rejected by most scientists today. Schrödinger was the first to give DNA a special status as an aperiodic crystal, but he surely did not mean it to be seen as a vital force.

Jacques Monod was completely against the concept of a vital

force, and a staunch defender of reductionism. Monod, of course, had shown that genes could respond to their environment through interactions of proteins, and at that time it was also known that enzymes could control their own activities in a similar way. For instance, once there is enough biochemical energy in your cells, a chemical signal goes to the enzymes that break down glucose to tell them to stop working. For Monod, this chemical 'intelligence' was proof of the power of reductionism. It had evolved through blind chance, as a result in DNA mutations that had served the organism well and so had survived. So Monod was scathing of those who thought there was any higher purpose or driving force to life – all could be explained by chemical bonding!

One essential theme in evolution is its 'blindness'. Mutation happens at random. Most mutations either do not affect, or are deleterious to, the organism. Occasionally they will give an organism an advantage in an environment, for example one that gives a mammal white fur in an Arctic habitat. But, given the Central Dogma, the animal cannot, while it is alive, 'decide' to have a gene for a white coat because it snows a great deal.

However, in 1988, John Cairns and his co-workers reported some experiments that seemed to suggest that bacteria, at least, have mechanisms for 'choosing' which mutations would occur. This is called directed mutation. In these experiments, bacteria were grown on a nutrient medium that lacked a vital component – such as the amino acid tryptophan. Mutations that conferred the ability to synthesise the missing tryptophan seemed to occur more often than chance would predict. This was interesting, because the question of whether bacteria, like the rest of the living world, mutated randomly was thought to have been settled by experiments carried out by Nobel Prizewinners Salvador Luria and Max Delbrück in the 1940s. Called the fluctuation test, the experiment looked at the acquisition of viral resistance in bacteria. The distribution of these mutants among the population seemed to suggest clearly that mutation was independent of environmental pressure – in this case the presence of viruses in the environment.

Cairns' work is interesting because it appears to revive the discredited theory of the French naturalist Jean Lamarck that characteristics acquired during an organism's lifetime could be inherited. Lamarck argued, for example, that the long neck of the giraffe was the result of generations reaching up for food. With the development of the Central Dogma, Lamarckism was discredited, as information cannot flow back from protein to DNA. Stretching up to tall trees, if you are a giraffe, has no effect on DNA; it does not cause a mutation in the DNA for neck muscles.

It is too early to say whether Cairns' experiments signal a revival of Lamarck's ideas. Acquired characteristics (the ability to make tryptophan because of deprivation during the life of a bacterium) would be passed on, just as Lamarck saw reaching to tall trees being passed down from one giraffe to its offspring. It could be something peculiar about bacteria, or about starvation generally (the lack of an essential amino acid is a form of starvation for these bacteria).

Genes and the environment

In the selfish gene theory the environment, whether the personal inner or outer surroundings of an individual, or the wider global environment, is a passive backdrop to the relentless egotism of DNA. But many scientists argue that this rigid application of the Central Dogma should be peripheral, rather than central, to our thinking.

In particular, with the Human Genome Project and the scramble to identify genes 'for' particular disorders, we may be in danger of oversimplifying the link between genotype and phenotype. The danger of such thinking, according to American biochemist Richard Strohman, is that the public may become unnecessarily alarmed, and large sums of money could be spent on useless screening programmes. Strohman argues that while the one gene one disease, linear mapping of genotype onto phenotype is perfectly valid for single gene disorders such as haemophilia, it is on less secure ground when looking at commoner conditions such as cancer, heart disease, and mental illness.

Instead, Strohman argues for a re-examination of the concept of epigenesis – the interaction of genes with physiological and environmental factors. With multigenic disorders, there is a great deal of redundancy in the system, and there may be many unpredictable pathways to the same physiological outcome. Obviously there is great excitement if a gene defect is linked with a disease. For instance, a mutation in the gene for angiotensin-converting enzyme (ACE) has been linked with heart attack (myocardial infarction) (see also Chapter 7). As this is a major killer in the Western world, there is obvious pressure to look for cause and effect. It is likely, however, that in the presence of a defective ACE gene, some other gene takes over. Therefore ACE mutation is possibly to be regarded as a necessary – but not sufficient – condition for development of heart attack. Screening the whole population, or even vulnerable families, may not, argues Strohman, be of any ultimate value although it will certainly consume precious resources.

Looking to the wider environment, the interaction of genes and their products with their surroundings may be a two-way process. Just as the presence of heavy metals in soil may signal the turning on and expression of genes for proteins that mop them up, so biological processes have an effect on their environment. These ideas are encapsulated in the theory of Gaia, developed by British chemist James Lovelock, as well as that of Lynn Margulis, whose work on endosymbiosis was discussed in Chapter 4.

According to Gaia theory, life and its environment are much more tightly coupled than was previously realised. Many of the substances essential to life, such as oxygen and carbon dioxide in the atmosphere, are actually produced by living things. Lovelock argues that life itself regulates and repairs the conditions necessary for it to exist. Note here we are talking about life, rather than human life. The human species may well wipe itself out – compared to archaebacteria we are not so robust – but life itself is likely to persist. Both Lovelock and Margulis sometimes anger environmentalists concerned with the state of the planet for their grandchildren, by pointing out that the emergence of oxygen on the biosphere would have been regarded as pollution in the same way as we regard the nitrogen oxides of car exhausts as pollution. Yet oxygen allowed 'advanced' organisms, such as ourselves, which depend on oxygen, to emerge.

The key feature in regulation of the environment for life is temperature. If we look at our neighbouring planets Venus and Mars, we can see that they are inhospitable. One is too cold; one is too hot. Earth – like Baby Bear's porridge in *Goldilocks and the Three Bears* – is just right. A simplistic view of this is that because we are in between hot Venus and cold Mars our temperature is at some convenient in-between value. This is not so, because the Sun is increasing in luminosity all the time and will, eventually, burn away completely, at which time all life on our planet will inevitably become extinct unless technology finds some ingenious way of existing without the Sun.

The temperature of the Earth should have increased, because of the increasing output of the Sun, yet it has stayed roughly the same – an average of around 15°C – since life began around 3.8 billion years ago. Lovelock attributes this to the activity of our very early ancestors, the cyanobacteria. Their massive photosynthetic activity, he says, pulled carbon dioxide out of the atmosphere and put a stop to a massive greenhouse effect that would have shrivelled life up.

Critics argue that Gaia – the name comes from the Greek goddess of the Earth and was given to the idea by the late novelist William Golding, a friend of Lovelock's – is not a new idea, and is not a testable theory. Lovelock points to recent evidence which shows that the temperature of the Earth can be regulated by the gas dimethylsulphide, which is oxidised in the atmosphere to tiny droplets of sulphuric acid. These act as so-called cloud condensation nuclei that cool the planet. Therefore the action of bacteria is regulating the temperature of the biosphere. Such mechanisms, which have been discovered only in the last few years, do seem to suggest that the interactions between living things and their environment have a significance far beyond that which is predicted by molecular biology.

It is important to realise that both Margulis and Lovelock still accept the overall concept of Darwinism, and neither is putting forward any teleological (life has some underlying purpose) view, or anthropic principle. But Margulis, as has been discussed, feels that symbiosis – where large chunks of genetic material are transferred – is more of a driving force behind evolution than is the random mutation so much emphasised by molecular biologists.

New frontiers – chaos and morphic resonance

There is something odd about the very existence of life in the sense that living things are systems with a high degree of organisation, and co-ordination. It is hard to see how the study of DNA can address this issue directly. Stepping back from biology for a moment, and looking at the laws of physics – and in particular at thermodynamics, the study of energy – life seems to go against the direction in which the Universe is headed. According to the second law of thermodynamics, the amount of entropy, or disorder, in the Universe is constantly on the increase. Spontaneously energy flows from hot to cold, disordering the molecules in the cold substance, so the ice in your gin and tonic melts rather than the whole drink freezing up. Overall, the drink ends up with less order and structure. The tension between the trend towards the 'heat death' of the Universe, and the build up of order and complexity has been explored by British physicist Paul Davies in his book *The Cosmic Blueprint*. DNA cannot tell us a great deal about complexity, but the study of complex systems has come into its own over the last decade or so.

As far as the study of nature is concerned, complexity – like the world of the cell – is the rule rather than the exception. Complex behaviour was thought to be too difficult to study using the laws of physics. Unknown to mainstream science, however, the mathematical tools that would open up the mysteries of complex systems were developed by French mathematician Henri Poincaré in the 1920s. It was only with the advent of very powerful computers that this branch of mathematics (topology) could be exploited to show how complexity can arise within a deterministic system. This study is known as chaos and is applicable to many systems – from the weather through to human physiology and the solar system.

Chaos theory may have much to teach us about the cells' and organisms' function. Stuart Kauffman of the University of Pennsylvania has attempted to forge links between chaos theory and genetics. He looks at the pattern of gene regulation and expression within a cell as if it were a simple electronic switching system. If the system is an organism's genome, and the elements in the system are genes, then the genes are like simple switches that can be either on or off. The pattern of on–off is characteristic of a given cell. For instance, in a brain cell the insulin gene is in the off state, but in a pancreatic cell it is in the on state. As we have seen, a number of factors – proteins and the presence of other genes – seem to determine whether a gene is on or off. Kauffman calls these 'inputs' and argues that according to mathematical theory the number of inputs to a gene produces a system that hovers between order and chaos, within which evolution can start to build up complexity. Kauffman sees the cell as an attractor – a state of gene expression to which the organism gravitates from the billions of possible patterns. From his theory, Kauffman has been able to make fascinating predictions about how cells differentiate into relatively few cell types, the relationship between the number of genes in an organism and the number of cell types. For a human, assuming 100 000 genes, 370 cell types are predicted. Currently the number stands at 254 – which is of the right order, and of course, more could be discovered.

Chaos theory may also shed some light on the origin of life. According to Nobel Prize winner Ilya Prigogine, the early Earth may have been what he calls an 'excitable medium' ripe for a sudden leap into complexity – perhaps RNA, perhaps the cell. Excitable media are much in vogue among scientists in this field; Brian Goodwin of the Open University argues that this concept could explain development in a wide range of organisms – from plant flowering to the unfolding of a human embryo.

The ideas of chaos theory have also been brought to bear on the physiological rhythms whose relationship to gene expression remains obscure. Heartbeat is healthier within the redundant mechanisms of chaos rather than the rigidity of a regular rhythm, for instance. Many anatomical systems such as the capillaries of the lungs and the blood vessels have a fractal geometry, which is characterised by self-similarity over many orders of magnitude.

The same patterns repeat themselves in many aspects of nature – from insect ecology to the turbulent movements of the weather. Gaia theory has found connections between living things and their inanimate environment. The challenge now is to see what links can be found between evolution and chaos theory, in the same way as physicists are trying to link quantum mechanics and gravity.

The search for some kind of grand unified theory – the attempt to find wide-ranging patterns in nature – is perhaps part of human psychology. One of the most controversial of these to emerge in recent years is that of British botanist Rupert Sheldrake. In his book *A New Science of Life* published in 1981, Sheldrake argues that the Universe is permeated by intangible phenomena known as morphic fields. These contain information about natural systems and are formed whenever some new system, such as a crystal or new species or a pattern of behaviour, comes into being. These fields apparently evolve through space and time and the resonance between current events such as crystals forming or lessons being learned or embryos developing is known as morphic resonance. Sheldrake argues that morphic resonance can fruitfully address key problems in development, and that its study in the 1940s was interrupted only by the development of molecular biology.

There is little experimental evidence for morphic resonance and Sheldrake's ideas have been sharply criticised by the scientific community – to the extent that his first book was denounced by the journal *Nature* as being fit only for burning. In an editorial, Sir John Maddox said that molecular biology could provide the answers for any outstanding problems in biology.

No-one knows if Maddox's reductionist views will be borne out by data from the Human Genome Project. Remembering that the genes of different species are being mapped in parallel raises the question of the close relationship between ourselves and other species. Approximately one per cent difference in DNA between us and our mammal neighbours, such as the chimpanzee and the mouse, has led to language, and consciousness – the abilities needed to develop genetic engineering and all the other DNA technologies. Powerful as it is, it is unlikely that DNA alone can explain why we have discovered its secrets and their potential to meet human needs!

Further reading

CHAPTER 1

Horace Freeland Judson, The *Eighth Day of Creation*, Touchstone Books, S & S Trade, 1980.
James D. Watson, The *Double Helix*, Norton, 1980.

CHAPTER 2

Walter J. Moore, *Schrödinger: Life and Thought*, Cambridge University Press, 1992.
Erwin Schrödinger, *What is Life?*, Cambridge University Press, 1968.

CHAPTER 3

Evelyn Fox Keller, *A Feeling for the Organism: The Life and Work of Barbara McClintock*, W. H. Freeman, 1984.

CHAPTER 4

A. G. Cairns Smith, *Seven Clues to the Origin of Life: A Scientific Detective Story*, Cambridge University Press, 1990.
Adrian Desmond and James Moore, *Darwin*, Warner, 1992.
Bernard Dixon, *Power Unseen: How Microbes Rule the World*, W. H. Freeman, 1994.
Lynn Margulis and Dorion Sagan, *A Garden of Microbial Delights: A Practical Guide to the Subvisible World*, Kendall Hunt, 1993.

John Postgate, *The Outer Reaches of Life*, Cambridge University Press, 1994.

CHAPTER 7

John Harris, *Wonderwoman and Superman: The Ethics of Human Biotechnology*, Oxford University Press, 1992.
David Wetherall, *The New Genetics and Clinical Practice*, 3rd edition, Oxford University Press, 1991.

CHAPTER 8

David W. E. Smith, *Human Longevity*, Oxford University Press, 1993.
Lewis Wolpert, *The Triumph of the Embryo*, Oxford University Press, 1994.

CHAPTER 10

Robert Walgate, *Miracle or Menace: Biotechnology and the Third World*, Panos Books, 1990.

CHAPTER 12

Paul Davies, *The Cosmic Blueprint*, Touchstone Books, S & S Trade, 1989.
Richard Dawkins, *The Selfish Gene*, 2nd edition, Oxford University Press, 1990.
Brian Goodwin, *How the Leopard Changed its Spots*, Weidenfeld, 1994.
Stuart Kauffman, *The Origins of Order: Self Organisation and Selection in Evolution*, Oxford University Press, 1993.
James Lovelock, *The Ages of Gaia: A Biography of our Living Earth*, Bantam, 1990.
Rupert Sheldrake, *The Presence of the Past: Morphic Resonance and the Habits of Nature*, Harper Collins, 1989.

Index